Nondestructive Techniques for the Assessment and Preservation of Historic Structures

Nondestructive Techniques for the Assessment and Preservation of Historic Structures

Edited By

Luisa M. S. Gonçalves, Hugo Rodrigues and Florindo Gaspar

CRC Press
Taylor & Francis Group
Boca Raton London New York

CRC Press is an imprint of the
Taylor & Francis Group, an **informa** business

MIX
Paper from responsible sources
FSC® C014174
FSC
www.fsc.org

CRC Press
Taylor & Francis Group
6000 Broken Sound Parkway NW, Suite 300
Boca Raton, FL 33487-2742

Printed on acid-free paper

International Standard Book Number-13: 978-1-138-71047-4 (Hardback)

Visit the Taylor & Francis Web site at
http://www.taylorandfrancis.com

and the CRC Press Web site at
http://www.crcpress.com

Printed and bound in the United States of America by Sheridan

Contents

Preface

The preservation of the built heritage has long been a public concern, mainly due to fears about the loss of identity, history and heritage of populations. The main concerns are the conservation and restoration of monuments that usually represent important events in the history of a city or a country. More recently, urban residents and policymakers have become aware of the abandonment or degradation of old city cores, leading to mischaracterisation of the buildings and ways of living.

To preserve history and promote building and urban renewal, considering the basic principles of the preservation process, it is important to introduce the least possible disturbance. To start with, the diagnostic process is a key aspect, especially to investigate the construction characteristics and the damage to materials, and to find structural and nonstructural problems.

To start any process, a visual inspection, a study and knowledge of the original construction methods and materials and historical repair techniques can help but may not be sufficient, and the use of conventional techniques to complement the information needed can result in an insufficient understanding or in extensive and unnecessary intrusions in the construction.

In recent years, the rapid growth of science and research, combined with the industry and the need to gather more and accurate information, have led to the fast development of nondestructive testing methodologies that allow the architectural archaeology to be studied, the structural assessment to be supported and information to be given about the material properties. Each technique can be used for a specific purpose, but, in some cases, only a combination of techniques is reliable and gives an accurate interpretation of the data acquired.

The fundamental contribution and aim of this book is to give a full overview of several case studies where different nondestructive techniques have been applied, in several cases using multidisciplinary approaches, which aim to highlight the importance of the information acquired and encourage the use of these techniques in future studies.

The book brings together 16 chapters focused on nondestructive testing techniques applied at the urban building level and also applied to monumental buildings, archaeology and cultural heritage, bringing together more than 40 international researchers and experts in the field, who are the source of practical case studies supported by a theoretical background.

Luisa M. S. Gonçalves, Hugo Rodrigues and Florindo Gaspar

Institutional Support and Acknowledgements

PARTNERS

POLITÉCNICO DE LEIRIA | ESCOLA SUPERIOR DE TECNOLOGIA E GESTÃO

POLOTSK STATE UNIVERSITY · 1581 · DOCENDO DISCIMUS

FUNDING

Co-funded by the Tempus Programme of the European Union

RETHINKe

REFORM OF EDUCATION THRU INTERNATIONAL KNOWLEDGE EXCHANGE

This publication has been funded with support from the European Commission within the framework of the Tempus Project Reform of Education THru INternational Knowledge Exchange, number: 544178-TEMPUS-1-2013-1-PT-TEMPUS-JPCR.

This publication reflects the views only of the authors, and the Commission cannot be held responsible for any use which may be made of the information contained therein.

Editors

Luisa M. S. Gonçalves holds a PhD in civil engineering in the scientific area of urban and regional planning, an MSc in geographic information systems and a specialization in building information modelling. She is a senior lecturer at the School of Technology and Management, Polytechnic Institute of Leiria where she is responsible for the Laboratory of GIS and Transports and was head of the Civil Engineering Department. She is also a researcher of INESC Coimbra and a visiting professor of the master's programme in science and geographic information systems at the NOVAIMS-New University of Lisbon, teaching cartographic sciences. She has been involved as a coordinator or team member in research projects, educational projects and specialised consultancy studies ordered by several public institutions and companies, all of which are related to the field of urban planning, remote sensing and geographic information systems. Her main research interests are related to 3D GIS modelling, uncertainty modelling and the application of multispectral image analysis to cultural heritage. She is an author of more than 50 scientific papers, published in peer-reviewed scientific journals and proceedings of international and national conferences. She was an active member of COST Action TU1208-Civil Engineering Applications of Ground Penetrating Radar and COST Action TU1206 SUB-URBAN – A European network to improve understanding and use of the ground beneath our cities.

Hugo Rodrigues is a senior lecturer at the civil engineering department of the School of Technology and Management, Polytechnic Institute of Leiria, Portugal, teaching several topics related to structural analysis, building pathology and rehabilitation. He received his PhD in Civil Engineering from the University of Aveiro in 2012. His experience is in seismic analysis, having participated as a team member in research and development projects, specialized consultancy studies ordered by several public institutions and companies regarding the assessment of seismic risk. His major research interests are building rehabilitation, structural health monitoring and seismic safety, including experimental and numerical activities. He is a research member of the RIsks and Sustainability in COnstruction (RISCO) research centre. He has co-authored more than 150 publications in top-tier peer-reviewed journals and national and international conferences.

Florindo Gaspar is senior lecturer at the Civil Engineering Department of the School of Technology and Management, Polytechnic Institute of Leiria (IPLeiria), where he is responsible for subjects related to wooden and masonry structures, materials and processes of construction, building installations. He was responsible for the Laboratory of Building Materials and coordinator of the graduation and master degree programs in civil engineering. He is currently a researcher at the Centre for Rapid and Sustainable Product Development (CDRsp). He has an MSc in construction and a PhD in civil engineering from the Instituto Superior Técnico (University of Lisbon), and a degree in civil engineering from the University of Coimbra. He has been involved in various consultancy studies ordered by several companies. He is the author of several scientific articles published in journals and national and international conferences. His research interests are related to monitoring, rehabilitation and conservation of wood and masonry structures, having supervised some MSc projects in this area.

Contributors

Andreia Almeida
School of Technology and Management
Polytechnic Institute of Leiria
Leiria, Portugal

António Arêde
CONSTRUCT-LESE
Faculty of Engineering
University of Porto
Porto, Portugal

Rafael Asorey-Cacheda
Defense University Center
Spanish Naval Academy
Marín, Spain

Sofia Avlonitou
Geoservice Ltd.
Athens, Greece

Aliaksandr Bakatovich
Civil Engineering Faculty
Polotsk State University
Novopolotsk, Belarus

Pedro Cardoso
School of Technology and Management
Polytechnic Institute of Leiria
Leiria, Portugal

Oriol Caselles
Universitat Politècncia de Catalunya
Barcelona, Spain

Tiago Cerejo
School of Technology and Management
Polytechnic Institute of Leiria
Leiria, Portugal

Nadezda Davydenko
Civil Engineering Faculty
Polotsk State University
Novopolotsk, Belarus

Klisthenis Dimitriadis
Geoservice Ltd.
Athens, Greece

Ana P. Falcão
CERIS – Civil Engineering Research and Innovation for Sustainability
Instituto Superior Técnico – Universidade de Lisboa
Lisbon, Portugal

Paulo Fernandes
CERIS, School of Technology and Management
Polytechnic Institute of Leiria
Leiria, Portugal

Anton Finogenov
Civil Engineering Faculty
Polotsk State University
Novopolotsk, Belarus

Florindo Gaspar
School of Technology and Management
Polytechnic Institute of Leiria
Leiria, Portugal

Tiago Gomes
School of Technology and Management
Polytechnic Institute of Leiria
Leiria, Portugal

Luisa M. S. Gonçalves
School of Technology and Management
Polytechnic Institute of Leiria
Leiria, Portugal
and
INESC COIMBRA
Nova IMS University
Lisbon, Portugal

Ramon González-Drigo
Resistència de Materials i Estructuresa l'Enginyeria
Universitat Politècncia de Catalunya
Barcelona Tech
Barcelona, Spain

Aliaksandr Ivanenko
Civil Engineering Faculty
Polotsk State University
Novopolotsk, Belarus

Ioannis Konstantakis
Geoservice Ltd.
Athens, Greece

Melda Küçükdemirci
Geophysical Engineering Department
Istanbul University
Istanbul, Turkey

Susana Lagüela
Department of Cartographic and Terrain Engineering
University of Salamanca
Ávila, Spain

Andrei Lazouski
Civil Engineering Faculty
Polotsk State University
Novopolotsk, Belarus

Dzmitry Lazouski
Civil Engineering Faculty
Polotsk State University
Novopolotsk, Belarus

Yahor Lazouski
Civil Engineering Faculty
Polotsk State University
Novopolotsk, Belarus

Vítor Lopes
School of Technology and Management
Polytechnic Institute of Leiria
Leiria, Portugal

Georgianna Moraitou
Hellenic Ministry of Culture
National Archaeological Museum
Department of Conservation
Physical–Chemical Research and Archaeometry
Athens, Greece

Lara Pajewski
Department of Information Engineering, Electronics and Telecommunications
Sapienza University of Rome
Rome, Italy

Ioannis Panagakos
Hellenic Ministry of Culture
National Archaeological Museum
Department of Conservation
Physical–Chemical Research and Archaeometry
Athens, Greece

Aliaksandr Paprauka
Civil Engineering Faculty
Polotsk State University
Novopolotsk, Belarus

Micael Pedroso
School of Technology and Management
Polytechnic Institute of Leiria
Leiria, Portugal

Vega Pérez-Gracia
Resistència de Materials i Estructuresa l'Enginyeria
Universitat Politècncia de Catalunya
Barcelona Tech
Barcelona, Spain

Patrick Pinheiro
School of Technology and Management
Polytechnic Institute of Leiria
Leiria, Portugal

Iván Puente
Defense University Center
Spanish Naval Academy
Marín, Spain

Belén Riveiro
Applied Geotechnologies Group
University of Vigo
Vigo, Spain

Hugo Rodrigues
RISCO, School of Technology and Management
Polytechnic Institute of Leiria
Leiria, Portugal

Javier Sanjurjo-Pinto
PhD Programme in Protection of the Cultural Heritage
University of Vigo
Vigo, Spain

Pedro Santos
Vesam Group
Leiria, Portugal

Sónia Santos-Assunçao
Murphy Surveys
Dublin, Ireland

Nelson Silva
School of Technology and Management
Polytechnic Institute of Leiria
Leiria, Portugal

Mercedes Solla
Defense University Center
Spanish Naval Academy
Marín, Spain

Michael Styllas
Geoservice Ltd.
Athens, Greece

Humberto Varum
CONSTRUCT-LESE
Faculty of Engineering
University of Porto
Porto, Portugal

Flávio Vazão
School of Technology and Management
Polytechnic Institute of Leiria
Leiria, Portugal

João Veludo
School of Technology and Management
Polytechnic Institute of Leiria
Leiria, Portugal

1 Nondestructive Techniques for the Assessment and Preservation of Existent Constructions

Luisa M. S. Gonçalves, Florindo Gaspar and Hugo Rodrigues

CONTENTS

1.1 INTRODUCTION

Interventions in historic constructions are one way of preserving the cultural history of people, a region or even a country. According to Violla (2012), investment policies and scientific research can act as a stimulus to communities, not only to protect physical assets but also to preserve construction methods, which include histories, and environments, thus supporting the sense of continuity and identity, which will lead to the promotion of a physical regeneration of monumental complexes, isolated buildings or city centres. Choosing the most appropriate way of intervening in a building should require a careful decision that always takes into account its historical significance, as well as bearing in mind a number of other concerns, such as its physical condition. That is, it is necessary to understand the pathologies or the degree of the integrity of the materials. In addition, monitoring the conservation state of historical constructions is extremely important for planning any maintenance interventions, thus allowing the associated costs to be minimised. However, construction health monitoring usually requires a wide variety and combination of procedures, including traditional methods and new technologies, which always require (1) exhaustive on-site visual inspection; (2) sample extraction for laboratory analysis; and (3) detailed and time-consuming analysis of the data (Valença et al. 2013). Due to recent developments in digital equipment, new opportunities are emerging for construction assessment. Thus, research studies have focused on the development of digital technology applications for the analysis of heritage degradation and degradation-monitoring surveys (Zheng 2015).

The present chapter gives an overview of some nondestructive technologies. It provides technical knowledge regarding data acquisition techniques that are used in the characterisation and preservation of important monuments and common constructions, such as multispectral images, geophysics data (ground-penetrating radar [GPR]), flat-jack tests, infrared thermographic images, laser scanning data and ultrasound, which will be applied and discussed in detail in each chapter. Despite the availability of detailed overview information on these nondestructive techniques (e.g. Wehr and Lohr 1999, Lillesand and Kiefer 2000, Li et al. 2008, Moropoulou et al. 2013), there are always new data regarding the rapid development and the new applications used in cultural heritage, on the one hand, and on the other hand, facing the challenge of being capable of shedding new light on these new interesting techniques and application examples. Thus, the various chapters of this book provide a wide range of case studies based on real-world experience using samples gathered from different geographical realities.

1.2 NONDESTRUCTIVE TECHNIQUES FOR DATA ACQUISITION AND PROCESSING

1.2.1 Terrestrial Laser Scanning

1.2.1.1 Short Overview

The development of laser scanning (LS), also referred to as LiDAR systems (light detection and ranging), and its use on different platforms (airborne laser scanning [ALS] or terrestrial laser scanning [TLS]) go back to the 1970s. Laser scanning has revolutionised 3D data acquisition, and in contrast to the 'classical' manual data acquisition techniques, such as terrestrial surveying and analytical photogrammetry, these automatic recording methods allow an automated dense sampling of the object surface within a short time (Pfeifer and Briese 2007). Laser scanning systems constitute active sensing systems, which use a laser beam as a sensing carrier (Wehr and Lohr 1999), and they measure the time lag between the emission of a laser beam and the detection of its backscattered echo. These systems generate geometric results in terms of distance, position, intensity and coordinates (Ackermann 1999). The emitted beam travels through the atmosphere and interacts with objects along its path. The strength of the backscatter may be recorded as well, resulting in what is normally referred to as 'intensity', a value that is given as a digital number. The 'intensity' values recorded by laser scanning measurements are subject to a wide range of influences, such as spreading loss, surface roughness, object reflectivity and atmospheric attenuation. Although these data have been mainly used for visualisation purposes, the expectation is that the normalisation of these measurements will be suitable for classification purposes (Pfeifer and Briese 2007).

Terrestrial laser scanners mainly use round trip time measurements and phase-based ranging. Round trip time ranging scanners are better suited for outdoor operations where longer ranges are measured. The purpose of using the largest ranges probed with the pulse round trip time measurement principle is to obtain centimetre accuracy. Shorter distances, for example up to 100 m, can be measured faster and more accurately with the phase-based measurement technique. Shorter distances, for example up to 2 m, can be measured with even higher precision, for example, accuracy better than ±1 mm, with triangulation (Pfeifer and Briese 2007).

The difference between pulse round trip time scanners and phase-based difference measurement scanners in TLS, according to Pfeifer and Briese (2007), is that pulse round trip laser scanners have a higher range, while phase-based laser scanners have higher measurement speeds and better precision. In a triangulating laser scanner, the laser energy is widened so as to form a plane rather than a beam. This method of scanning is restricted in depth, because the quality of the intersection of a bundle of rays has a tendency to decrease with range. Therefore, this type of scanner is restricted to ranges of one or even a few metres. The precision is normally better than ±1 mm. Further information about this method can be found in Pfeifer and Briese (2007).

With the rise of low-cost 3D, imaging systems can be seen as an economic and available simple imaging alternative. These systems have the advantage of having much lower investment costs and are usually very user-friendly (Paulus 2014). In addition, they can be used without intensive training. However, common low-cost sensors have the limitation of a trade-off between range and accuracy, which provides either a low resolution of single objects or a limited imaging field.

For most applications, a single sensor is commonly not able to fulfil all the requirements. Thus, the use of multiple sensors is necessary (Dupuis et al. 2014). The advantages and potentials of the combined use of multiple sensors have been presented in Hall and Llinas (1997) and are also used nowadays in cultural heritage (Guidi et al. 2008). Dupuis et al. (2014) have recently developed an automated approach focusing on the fusion of unorganised point clouds that came from a different sensor using a characteristic 3D shape. The approach combines the advantages of using multiple sensors when measuring volume and resolution, and this leads to generating a multi-resolution map.

More information about the TLS technique can also be found in Chapters 12 and 14 of this book.

1.2.1.2 Applications of Terrestrial Laser Scanning Data

The field of applications for TLS and ALS is very diverse. Several studies have demonstrated that TLS can be considered an interesting option for acquiring a digital terrain model, archaeological site model or city model including the roof forms and the geometry of the facades (e.g. Ackermann 1999, Guidi et al. 2008, Lerma et al. 2010). Usually, imagery obtained with a digital camera is also acquired, so that the geometric model can be textured. This combination enables an efficient, flexible and reliable acquisition, with the advantage of a photorealistic representation, and is mostly used in terms of the documentation of cultural heritage (Dupuis et al. 2014). In the documentation of cultural heritage, laser scanning is applied, ranging from small artefacts of a few centimetres in diameter, to statues and monuments with a height of one or a few metres, up to entire buildings and castles (e.g. Guidi et al. 2008). The accuracy with which this can be achieved is dependent on the relationship between the LS point-sampling density and the size of the objects to be classified or modelled (Steel et al. 2003). Since the precision and measurement volume requirements are very diverse, more than one laser scanner is normally used to cover all the applications in cultural heritage. Figure 1.1 shows an example of scans acquired with two different TLS systems to obtain a 3D model of a thumb in the Santa Maria da Vitoria Monastery of Batalha.

Recent applications also show that this technology can be used not only to record the object geometry and play a role in record-keeping aspects, but also to protect against targeted degradation and to provide a more meaningful interpretation function (Zheng 2015). In Chapter 9, TLS was used to verify the verticality of the Founder's Chapel of Batalha's Monastery, and in Chapter 12, examples are presented of TLS application to ancient masonry arch bridges to obtain a quicker measurement of the visible geometry of the bridge, including its superficial pathologies, giving extra information regarding their sizes and depths, which are the latest indicators of their severity.

The integration of TLS with other nondestructive methods has also been used to improve the comprehension of interactions between different phenomena and to allow a better decision and planning process. In Chapter 14, a 3D reconstruction of the Roman site *Aquis Querquennis* (Bande, Spain) was performed using TLS data acquisition combined with 3D GPR data acquisition. The data were then georeferenced, and the hybrid outcomes were merged into a single image, allowing the reconstruction of the visible reality of the archaeological site with the integration of the unexcavated subsurface structures that were detected and their integration into a geographic information system (GIS).

FIGURE 1.1 Two scans performed in the Santa Maria da Vitoria Monastery of Batalha, Portugal, which were acquired with two different TLS systems.

1.2.2 Remote Sensing

1.2.2.1 Short Overview

Remote sensing is the science of obtaining information about an object, area or phenomenon through the analysis of data acquired by a sensor that is not in contact with the object, area or phenomenon (Lillesand and Kiefer 2000). However, remotely collected data can be in many forms, and in this section, only a short overview is presented of the electromagnetic energy data recorded by sensors that operate from drones, airborne or space-borne platforms, or even a terrestrial camera and their application for the evaluation and documentation of heritage structures. These sensors can be active (generating their own source of illumination) or passive (measuring reflected sunlight), and they record variations in the way the surface features of an object reflect and emit electromagnetic energy. Electromagnetic energy has many forms. In applications to assist in inventorying, mapping and monitoring heritage constructions, visible light, near-red, thermal and microwave energy are usually used.

To automatically extract information from the multispectral images provided by passive sensors, there are several classification methods, which use different approaches and methodologies. Classification methods are based on the fact that different objects have different spectral characteristics, and the wavelength is dependent on their reflective characteristics. However, many materials have similar spectral characteristics in one wavelength range and quite different ones in others (Lerma 2005). This allows the selection of a wavelength range that will enhance the characteristics of a surface to detect damage.

The most commonly used classifiers are supervised ones, which perform the classification in two phases: 1) training areas, which are descriptors of the spectral characteristics of the classes, are identified for each class in the image; 2) the results obtained in the first phase are used, so that one or several classes (if soft classifiers are used) are associated with each spatial unit of the image. Another important stage in the production process is the validation of the results, which is usually done when comparing the classification results with the reference data that are obtained from a sample of points, and using field observations or other sources of higher accuracy, which are supposed to represent the ground truth.

In the described classification process, there are many sources of uncertainty that influence the final results of the classification. Therefore, the obtained results depend on the choices made and the approaches used, such as the training sample chosen and the applied classification method (Doan and Foody 2007).

1.2.2.2 Applications of Remote Sensing

Since the launch of the IKONOS satellite, which was the first to provide very high-spatial resolution data, the increasing availability of new active and passive satellite sensors that provide very high spatial resolution data has opened up new application opportunities to document and evaluate heritage constructions (Agapiou 2017).

Remote sensing archaeological research, for example, has been increasingly motivated during the last decades, and especially after the launch of the IKONOS satellite sensor in 1999 (Agapiou 2017). Recently, Nebbia et al. (2016) used remote sensing and GIS to create a risk map of Libyan archaeological sites.

Some methods were also developed to obtain accurate information on construction materials and the pathology of building facades and roofs. This was carried out through an automatic classification of visible and near-infrared photographs (Lerma 2005, Valença et al. 2013) and the gathering of information on roofing materials and pathologies, which was achieved through an automatic classification of visible and near-infrared aerial images (Gonçalves et al. 2009). There are also algorithms capable of detecting discontinuities in multispectral images, such as cracks that were analysed through a series of variations of the intensity of the pixels (Gonzalez and Woods 2002, Dare et al. 2012). Since cracks can be regarded as discontinuities in an image texture, their characterisation can be automatically found (Dare et al. 2012, Valença et al. 2013).

Multi-scale and multi-temporal approaches have also been developed based on the use of interferometric synthetic aperture radar (InSAR), particularly persistent scatterer interferometry (PSI), to detect and map the effects visible at the surface of both vertical and horizontal urban dynamics, including the susceptibility of heritage assets to structural damage (Tapete and Cigna 2012, Pratesi et al. 2016). These kinds of images, obtained by active sensors, and approaches have been shown to be particularly useful in historic urban centres, where these dynamics typically add onto existing issues related to structural and/or geological instability, the maintenance of the built heritage and the management of infrastructures.

In Chapter 3, an application of multispectral images of very large resolution, taken at different dates, was made for the automatic detection of changes over time of buildings' roofing materials and nonstructural anomalies.

1.2.3 Ground-Penetrating Radar

GPR is an electromagnetic pulse reflection method based on physical principles similar to those of seismic reflection. It is a geophysical technique for shallow investigations with high resolution, with several developments and new applications in the last decades. These geophysical methods are based on the emission of a very short electromagnetic pulse in a specific frequency band.

Frequencies between 10 and 1000 MHz are used for geological and engineering investigations, and frequencies higher than 1000 MHz are usually advantageous for material testing.

Usually, for monuments, the centre frequency of the antennae used is between 200 MHz and 1 GHz. The high-frequency antenna (1 GHz) is suited for very detailed and surface studies, and can be a useful tool for structure detail characterisation; it is used in the detection of cracks, estimation of wall thickness, and detection of humidity inside structures, different elements, reinforcing steel bars, and so on. For subsoil investigation, depending on the objective of the use, in the first 2–3 m, a 500 MHz antenna can produce a very high-resolution map. However, for higher depths, lower-frequency antennae are more useful (Arias et al. 2007).

The antenna is moved over the area under study, and an image of the shallow subsurface under the alignment used is obtained. The results are images representing the reflections detected, which are representative of the two-way travel time of the pulse emitted.

For the particular case of building rehabilitation, the investigation of the characteristics of building materials, changes to the original construction, the foundation, and the pathologies and characteristics of the sub-soil are usually required. For this, it is necessary to use nondestructive techniques, among which GPR can be significant as a result of its ease of implementation and reliability of results.

More information about GPR techniques can be found in the literature (Blindow et al. 2007, Conyers 2013, Annan 2009) and also in Chapters 12 and 15 of this book.

1.2.3.1 Applications of Ground-Penetrating Radar

Regarding the employment of GPR techniques for the evaluation of monuments and existent buildings, several cases can be found in the literature. In the key points observed in the majority of the works, a multidisciplinary approach has been used, combining LS, GPR, acoustic, electrical and radar methods supporting the finite element analysis in a final stage.

In monuments, GPR techniques are used in the geometric survey to estimate homogeneity or heterogeneity (Arias et al. 2007, Leckebusch 2000). Ranalli et al. (2004) used the GPR surveys in a medieval church located in L'Aquia with a view to planning the restoration, aiming at the collection of data on wall thickness, internal masonry structure and the location of detachments or cracks. The campaign allows the more degraded zones to be detected and reveals variations in wall thickness, enabling these data to be taken into account in the structural assessment of the monument. Costa et al. (2016) performed the calibration of a numerical model of a stone masonry arch railway bridge using dynamic modal parameters estimated from an ambient vibration test. To define the bridge geometry, GPR tests were performed to obtain the profiles of outer facings of the bridge located in abutments, piers and spandrel walls.

Barraca et al. (2016) used GPR techniques in a masonry house to investigate information for the rehabilitation process (namely, the local geological conditions), map infrastructure networks, locate removed or altered elements, understand modifications and changes in the construction materials, and characterise pathologies and fracturing. The main issues found were related to the choice of antenna, detecting the signatures of different materials, among other things, due to the lack of past experiments with controlled walls.

The use of GPR for the assessment of existent constructions can be a powerful technique. However, it requires a strong background of experience to help in the correct interpretation of the results obtained (Binda et al. 1998). The application of the GPR technique in monuments as well as regular constructions for the detection of pathologies and defects, voids, internal structural elements and a connection between elements, among other things, can encourage the regular use of these techniques not only for research purposes but also for practical architectural and engineering applications. Thus, several application examples are presented in this book, including applications for the inspection and characterisation of ancient masonry arch bridges (Chapters 12 and 13), archaeological prospecting (Chapters 13 through 15), and the assessment of monuments and historical buildings (Chapters 11 and 12), specifically the assessment of masonry columns and wooden beams (Chapter 8).

1.2.4 Infrared Thermography

Thermography consists of the colour mapping of a particular body or object referencing the different temperatures. Infrared radiation is emitted by the body's surface, and an infrared camera receives the information and converts it into a colour scale. The fundamentals of this technique are given in Chapter 12.

Passive thermography corresponds to the use of natural heat sources, such as solar radiation or the slowly varying temperature of a microclimate, whereas active thermography uses a noncontact thermal input onto a surface of the inspected body by means of lamps, hot or cold air guns, or devices making the surface vibrate. The temperature of the surface around the excited area is measured by infrared thermography, either during or after the application of a thermal input at a suitable frequency (Grinzato 2012). The thermal input can be applied using pulse thermography, performed by stimulating the object with a heating pulse and monitoring its surface temperature evolution during the transient heating or cooling phase, or with a lock-in thermography technique using thermal waves and resulting in sinusoidal temperature modulation (Carlomagno et al. 2005). Sfarra et al. (2016) proposed an innovative hybrid approach combining both the time component and the solar source to obtain quantitative information, such as the depth of the defect. This approach could be useful for the estimation of the depth of small defects that appear inside vertical known structures.

Thermography is used not only for construction surveys but also in other fields, such as mechanics, electronics, and the chemical, aeronautical, medical and military areas (Balaras and Argiriou 2002). In civil engineering, it is useful to distinguish different types of materials as well as heterogeneity, cavity detection, moisture and the degradation state.

Several works have used thermography for heritage inspection and conservation. Passive thermography was used by Grinzato et al. (2002b) to monitor the Scrovegni Chapel (Padova, Italy), to easily detect hidden structures made of different materials buried in the walls and to identify thermal bridges. Pantoja (2016) analysed the architectural complex of Mercedários (Belém do Pará, Brazil). The thermograms generated by the infrared camera allowed the traces of the transformations to which the building had been subjected, such as alterations of forms and closures of openings, in addition to hidden structures in the edifice, to be verified. Freitas (2016) used this technique to identify areas that had lower temperatures due to the presence of water evaporation on the surface at the Diocesan Museum of Mantua (Italy). Avdelidis and Moropoulou (2004) concluded that the recording of thermal images of the surfaces under investigation provided significant information for the assessment of the materials and techniques used in protecting cultural heritage. In particular, it can be used efficiently for the assessment of conservation materials and techniques on the subject of surface cleaning, stone consolidation and the restoration of masonry with repair mortars. Passive thermography was used in the survey and characterisation of the Founder's Chapel pathologies at the Monastery of Batalha, as described in Chapter 9.

Active thermography was used by Więcek and Poksińska (2006) to show the invisible structure under the gothic plaster work at the Castle in Malbork (Poland). In the study by Cennamo (2013), defects of mechanical relevance were detected, such as cracks and crevices at the bell tower of Sagittarius and the grange of Ventrile, located in the province of Potenza (Italy). Merla et al. (2009) applied this technique to the historical cultural heritage damaged in the L'Aquila earthquake for crack detection in buildings after the earthquake and moisture detection in frescoes. This technique has proved itself capable of helping to find detached tiles in mosaics at the Archaeological Museum of Naples and the degradation of plaster and frescoes at the archaeological site of Pompeii (Meola 2013). At the historical Arsenal of Venice, active thermography has been applied successfully to enable the understanding of wall bonding, moisture mapping, and the measurement of the thermal diffusivity of bricks and plaster to take place (Grinzato et al. 2002a).

Thermography has also been used in combination with other techniques. Different techniques, such as infrared thermography, ultrasonics and electric-type geophysical methods, were analysed to acquire information on the synergic use of the different methods, which may be useful for the estimation of the sources of the buildings' degradation (Meola et al. 2005, Carlomagno et al. 2005).

1.2.5 Flat-Jack Tests

1.2.5.1 Short Overview

One of the main aims of the nondestructive tests is to study the mechanical characteristics of the materials, and this acquires significant importance when the objective is to characterise the mechanics of old construction materials, and in particular masonry structural elements, due to the variability of the materials and their heterogeneity, combining materials with different properties (Binda and Tiraboschi 1999). There are several ways to characterise the material properties, but in many cases, samples cannot be removed for laboratory testing due to the dimensions of the elements and the historical value of the construction under study. The development of nondestructive mechanical testing techniques can play an important role in characterising the properties of materials, and the flat-jack testing technique has been used to accomplish this objective. With this test method, it is possible to determine in situ the properties of undisturbed masonry specimens of large dimensions with minor, controlled and reparable damage. The basis of the flat-jack testing was primarily applied in the field of rock mechanics (Rocha et al. 1966) and later adapted by Rossi (Rossi 1982) for use with masonry. Two tests can be performed with the same technique: the single flat-jack test and the double flat-jack test (Gregorczyk and Lourenço 2000).

The single flat-jack test is used for the assessment of the in situ compressive stress in existing structures (Figure 1.2a). After the cut is done, the installed compressive stress causes the masonry above and below the cut to get closer. The flat-jack will introduce pressure until the original distance between points above and below the cut is re-established. The existent compressive stress in the masonry is estimated from the flat-jack pressure multiplied by factors that account for the ratio of the bearing area of the jack in contact with the masonry to the bearing area of the cut, and the physical characteristics of the jack.

Another possibility is to perform a double flat-jack (Figure 1.2b) after the single flat-jack tests, which allows the stress-strain behaviour under compressions of the masonry wall to be assessed. The test is carried out by inserting two flat jacks into parallel cuts, and the flat-jack pressure is gradually increased, with a monotonic or cyclic load introducing a compressive stress on the masonry between the cuts.

1.2.5.2 Applications of Flat-Jack Tests

The flat-jack test provides a relatively nondestructive way of determining the in situ mechanical properties of masonry. This testing technique has been successfully used in regular brick and stone masonry structures, but its practice on rubble stone masonry structures is not so common. In these cases, the experimental procedures have to be adapted and calibrated (Binda and Tiraboschi 1999).

These testing methods are described by the American Society for Testing and Materials (ASTM 1991) and the International Union of Laboratories and Experts in Construction Materials, Systems and Structures (Reunion Internationale des Laboratoires et Experts des Materiaux, Systemes de Construction et Ouvrages) (RILEM 2004).

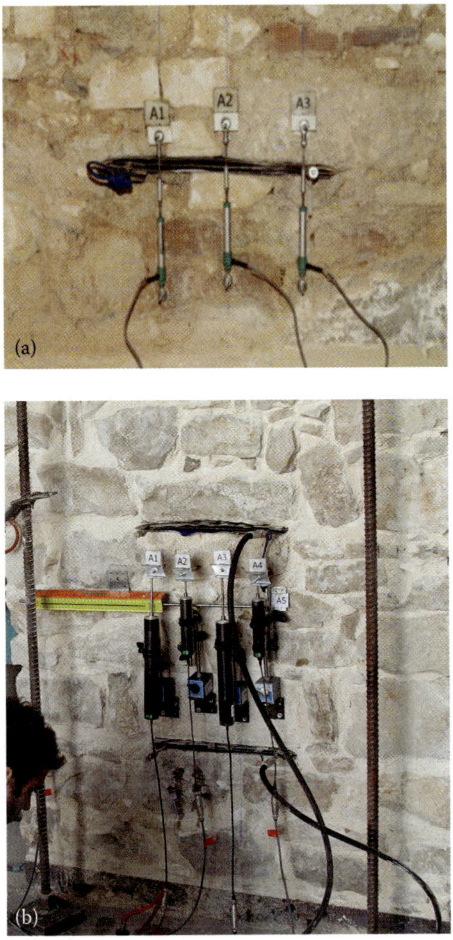

FIGURE 1.2 (a) Single flat-jack test; (b) double flat-jack test.

The flat jacks have been used in two main applications in terms of monuments: aiming to use the results for either a particular study of a building or an element of a monument (Binda et al. 2003, Carpinteri et al. 2005). In these cases, the flat jacks have also been combined with other nondestructive techniques, in particular with acoustic emission. Although it is difficult to obtain an easy direct relationship with other nondestructive techniques, the combined use increases the accuracy of the obtained results (Carpinteri and Lacidogna 2005, Binda et al. 2003).

Another case of a flat-jack application is its use to characterise the main properties of masonry walls in urban city centres, as in the case studies of Coimbra (Vicente et al. 2015a), Lisbon (Simões et al. 2015) and L'Aquila (da Porto et al. 2013, D'Ayala and Paganoni 2011). In the case presented in Chapter 4, single and double flat-jack tests were applied in four sites located in the city centre of Leiria, giving average results about the mechanical properties of the masonry infill, but also showing the heterogeneities found in the different buildings. In these cases, it is recognised that a large number of flat-jack tests are required to obtain reliable results regarding mechanical properties such as the maximum compressive strength, elastic modulus, which can be linked with the morphological and constitutional features of masonry walls and the state of conservation, so as to coherently establish the correlations among all of them (Vicente et al. 2015b).

1.2.6 Ultrasound

The purpose of this technique is to relate the properties of a material with the velocity of propagation of the ultrasound. Through the emission of ultrasonic pulses in the material and its reception at a certain distance, the speed of the waves crossing between these two points (emission/reception) is obtained by the following equation:

$$V = \frac{L}{T} \tag{1.1}$$

where:

V (m/s) is the ultrasound propagation velocity
L (m) is the distance between the transducers
T (s) is the transfer time of the ultrasound between the transducers indicated by the equipment

Three basic types of stress waves are created: longitudinal, shear and surface waves. Generally, the propagation is the fastest for longitudinal waves and the slowest for surface waves (Meola et al. 2005).

The use of ultrasound allows an efficient nondestructive evaluation of various materials, and it can be very useful in determining their characteristics and state of conservation.

The velocity variation is due to the density change. The more the density increases, the higher the ultrasound velocity will be.

Usually, ultrasound measuring devices consist of two transducers: one of them will have the function of transmitting and the other of receiving. There are three methods of performing the test: direct, semi-direct and indirect (Coelho 2010).

In the case of stone characterisation, it is possible, through the variation of the ultrasound velocity, to detect different values due to cracks, porosities, grain reorganisation or degradation due to weathering damage (Moreira 2013) or to make a quality classification (Kahraman et al. 2007).

Several works have used ultrasound for heritage inspection and conservation. Moreira (2013) performed ultrasonic tests on granite stone for the determination of some mechanical properties at the church of Nossa Senhora da Conceição (Porto, Portugal). Bromblet et al. (2012) performed long-term monitoring of the degradation of stone columns in the Church of Saint-Trophime of Arles in France. The first investigation took place in 1993 and the second in 2009. The velocity propagation results of the two campaigns, separated by a 16 year period, were compared, and the columns were separated into five classes of damage. Köhler et al. (1996) used ultrasound to analyse the stone material of the west portal of St Stephen's Cathedral in Vienna. By combining visual inspection with ultrasonic transmission measurements, all the stone elements could be clearly defined in terms of their lithotype. Binda and Saisi (2001) used ultrasonic tests and flat-jack tests together with other minor tests for the study of three similarly damaged cases of Italian church pillars: the Cathedral and the SS. Crocifisso Church in Noto, and the S. Nicolò l'Arena Church in Catania. Grossi and Lama (2015) used ultrasound velocity measurements to assess the state of the stone conservation of the Monument to Ramos de Azevedo (São Paulo, Brazil). Tavukçuoğlu et al. (2010) aimed to discover the thermal behaviour and ultrasonic characteristics of cracks in relation to the depth and the moisture content of the Cenabi Ahmet Paşa Camii, a sixteenth-century Ottoman mosque in Ankara. In Chapter 9 of this book, ultrasound was used in the survey and characterisation of the Founder's Chapel stone at the Monastery of Batalha. Ultrasound has also been widely applied in the nondestructive evaluation of wood (Bucur 2006) to evaluate decay, helping to map the damaged areas (Carrasco and Teixeira 2012), density and mechanical properties (Liñán and Hita 1995).

Several examples can be found of the use of ultrasound for heritage inspection and the conservation of wooden elements. Gatto et al. (2012) aimed to detect the external degree of decay of the wood *Araucaria angustifolia* (Bert.) O. Kuntze species, used in flooring in the historic building of Pelotas (Brazil). Conde et al. (2014) assessed the conservation status of the wooden frame of a building of great historic value known as 'El Corral del Conde' located in Seville (Spain), evaluating the compression strength and density loss due to deterioration of the elements. Estimation of density loss and strength was also done at the Ntra. Sra. de la Granada church (Spain) (Conde et al. 2013). Liñán et al. (2004) obtained the state of damage and the bending strength capacity of the timber roof of the Monastery of Santa Clara in Carmona (Seville). Kandemir-Yucel et al. (2007) applied this technique to timber samples on a thirteenth-century monument, Aslanhane Mosque in Ankara (Turkey), to find changes in moisture content due to microclimatic conditions. Liñán et al. (2015) used ultrasound and thermography techniques for the on-site inspection of wooden structures and the evaluation of their condition in a twentieth-century church in the town of Isla Cristina (Spain). Thermography identifies different materials and moisture content, while ultrasound detects the various degrees of deterioration and loss in density in areas of the wood with high moisture content.

1.2.5 Building Surveying and Appraisal

1.2.5.1 Short Overview

Building surveying and appraisal is an essential tool for monitoring the behaviour of a construction, allowing preventative maintenance, repair and rehabilitation.

It is fundamental that a specialist assesses in detail the conservation state of a construction, looking for problems or requirements of the structural and nonstructural elements that describe the actual condition of a construction in relation to its use, using the past, the actual use and future demands as a decision-making tool for future interventions and use of the construction.

Whatever the reason for evaluating or assessing the construction, these actions represent cost and time, and the results should be important in the decision-making process.

There are several ways of characterising and classifying the surveys and appraisals. One important aspect is the nature of the survey, whether it is used for continuous assessment, assessing the construction conditions along with the construction life cycle, or monitoring the changes with time to support maintenance and conservation work. The survey can also be related to a specific moment of the construction, capturing the image and state of conservation and showing the state at a specific moment in time of the construction.

Another use of surveying and appraisals can be applied to a large number of constructions. These surveys are used to assess the state of conservation of building stock, for example, in an old city centre, to support urban planning and to define rehabilitation policies. These surveys are usually performed using generic inspection sheets with checklists for the different construction parts. A significant volume of information is gathered, which is usually difficult to work with. The design of the database to work with this information needs to be carefully prepared to enable significant information, useful for the assessment and decision process, to be extracted.

1.2.5.2 Applications

Building surveying and appraisals have been widely used for research and in practical cases, for rehabilitation, conservation and/or strengthening a building, as well as characterising the state of existent constructions. In fact, the application of any nondestructive test should only be performed after a careful visual survey of the building or site conditions (Kordatos et al. 2013, Moropoulou et al. 2013).

Two different examples can be given for different types of structures. For the assessment of historical buildings, a preliminary on-site visual survey is performed to observe and record the prominent features of the building and of its immediate surroundings, converted into a detailed map locating areas of potential risk (Tavukçuoğlu et al. 2005). Similar procedures can be found for recent structures, where the results of the nondestructive test methods were used, for example, to reinforce a concrete bridge, so as to verify the preliminary conclusions of the visual survey (Choi et al. 2016), among many other examples in literature.

Visual surveys can also be applied after catastrophic events, such as earthquakes, for a construction or on a large scale. In these cases, visual damage interpretation allows the interpretation of the different behaviours of different types of structures and assists in defining guidelines for future construction practices for the country and surrounding regions (Gautam et al. 2016, Romão et al. 2013, Dolce et al. 2006).

Finally, and due to the urban regeneration of several historical city centres, the use of building surveys to characterise the global building stock has a strategic role in supporting the decision and implementation of the construction process. Several examples can be found in the literature, and in Chapter 2, the results of a survey of the historical city centre of Leiria are presented, with two levels of inspection, a preliminary inspection and a detailed inspection, which enable a clear picture to be formed of the main architectural characteristics of the city centre, the types of buildings, their uses, the state of conservation and the main pathologies. Similar studies have been promoted not only in other cities in Portugal (Santos et al. 2013, Vicente et al. 2015a, Lourenço et al. 2006, Furtado et al. 2016) but also in other cities around the word (da Porto et al. 2013, Theodoridou et al. 2011), which highlight the differences in the construction technologies and the main pathologies observed for the same countries and for other regions. Also, the surveys and characterisation of the building stock can be used to define and study other issues, such as energy efficiency, on an urban scale.

ACKNOWLEDGEMENTS

This work has been supported by the Fundação para a Ciência e a Tecnologia (FCT) under project grant UID/MULTI/00308/2013.

REFERENCES

Ackermann, F. 1999. Airborne laser scanning: Present status and future expectations. *ISPRS Journal of Photogrammetry & Remote Sensing* 54: 64–67.

Agapiou, A. 2017. Remote sensing heritage in a petabyte-scale: Satellite data and heritage earth engine© applications. *International Journal of Digital Earth* 10 (1): 85–102.

Annan, A. P. 2009. Electromagnetic principles of ground penetrating radar. In *Ground Penetrating Radar Theory and Applications*, ed. H. Jol. Elsevier. 1, 3–40, http://www.sciencedirect.com/science/article/pii/B9780444533487000016.

Arias, P., J. Armesto, D. Di-Capua, R. González-Drigo, H. Lorenzo, and V. Pérez-Gracia. 2007. Digital photogrammetry, GPR and computational analysis of structural damages in a mediaeval bridge. *Engineering Failure Analysis* 14: 1444–1457.

ASTM. 1991. Standard test method for in situ measurement of masonry deformability properties using the flat-jack method. In *ASTM C1197-91*. Philadelphia.

Avdelidis, N. P. and A. Moropoulou. 2004. Applications of infrared thermography for the investigation of historic structures. *Journal of Culture and Heritage* 5: 119–127.

Balaras, C. A. and A. A. Argiriou. 2002. Infrared thermography for building diagnostics. *Energy and Buildings* 34 (2): 171–183.

Barraca, N., M. Almeida, H. Varum, F. Almeida, and M. S. Matias. 2016. A case study of the use of GPR for rehabilitation of a classified art deco building: The InovaDomus house. *Journal of Applied Geophysics* 127: 1–13.

Binda, L., G. Lenzi, and A. Saisi. 1998. NDE of masonry structures: Use of radar tests for the characterisation of stone masonries. *NDT & E International* 31: 411–419.

Binda, L. and A. Saisi. 2001. Non destructive testing applied to historic buildings: The case of some Sicilian churches. In *Historical Constructions*, ed. P. B. Lourenço and P. Roca. Guimarães.

Binda, L., A. Saisi, and L. Zanzi. 2003. Sonic tomography and flat-jack tests as complementary investigation procedures for the stone pillars of the temple of S. Nicolò l'Arena (Italy). *NDT & E International* 36: 215–227.

Binda, L. and C. Tiraboschi. 1999. Flat-jack test as a slightly destructive technique for the diagnosis of brick and stone masonry structures. *Proceedings of the 8th International Conference on Structural Faults and Repair*, London, 13–15 July.

Blindow, N., D. Eisenburger, B. Illich, H. Petzold, and T. Richter. 2007. Ground penetrating radar. In *Environmental Geology: Handbook of Field Methods and Case Studies*, ed. K. Knödel, G. Lange, and H.-J. Voigt. Springer.

Bromblet, P., V. Vergès-Belmin, and S. Simon. 2012. Ultrasonic velocity measurements for the long-term monitoring of the degradation of marble columns in the cloister of the church of Saint-Trophime in Arles (France). *12th International Congress on the Deterioration and Conservation of Stone Columbia University*, New York: Columbia University.

Bucur, V. 2006. *Acoustics of Wood*. Berlin: Springer.

Carlomagno, G. M., R. di Maio, C. Meola, and N. Roberti. 2005. Infrared thermography and geophysical techniques in cultural heritage conservation. *Quantitative InfraRed Thermography Journal* 2: 5–24.

Carpinteri, A., S. Invernizzi, and G. Lacidogna. 2005. In situ damage assessment and nonlinear modelling of a historical masonry tower. *Engineering Structures* 27: 387–395.

Carpinteri, A. and G. Lacidogna. 2005. Damage evaluation of medieval towers using the acoustic emission technique. *WIT Transactions on the Built Environment* 83: 557–566.

Carrasco, E. and A. Teixeira. 2012. Methodology for inspection of wood pathologies using ultrasonic pulses. *CERNE* 18: 3.

Cennamo, G. M. 2013. The contribution of diagnostics in architectural survey; case study of combined thermography application. In *XI Forum internazionale di studi*. Engineering Faculty, International Telematic University UNINETTUNO, Rome, Italy.

Choi, W-C., M. Picornell, and S. Hamoush. 2016. Performance of 90-year-old concrete in a historical structure. *Construction and Building Materials* 105: 595–602.

Coelho, R. 2010. Metodologia geral para o restauro de construções históricas. MSc diss., Universidade de Trás-os-Montes e Alto Douro, Portugal.

Conde, M. J., C. Liñán, and P. Hita. 2013. Application of non-destructive techniques in the inspection of the wooden roof of historic buildings: A case study. *Advanced Materials Research* 778: 233–242.

Conde, M.J., C. Liñán, and P. Hita. 2014. Use of ultrasound as a nondestructive evaluation technique for sustainable interventions on wooden structures. *Building and Environment* 82: 247–257.

Conyers, L. B. 2013. *Ground-Penetrating Radar for Archaeology*. Altamira Press.

Costa, C., D. Ribeiro, P. Jorge, R. Silva, A. Arêde, and R. Calçada. 2016. Calibration of the numerical model of a stone masonry railway bridge based on experimentally identified modal parameters. *Engineering Structures* 123: 354–371.

Dall'O', G., A. Galante, and M. Torri. 2012. A methodology for the energy performance classification of residential building stock on an urban scale. *Energy and Buildings* 48: 211–219.

da Porto, F., M. Munari, A. Prota, and C. Modena. 2013. Analysis and repair of clustered buildings: Case study of a block in the historic city centre of L'Aquila (Central Italy). *Construction and Building Materials* 38: 1221–1237.

Dare, P., H. Hanley, C. Fraser, B. Ridel, and W. Niemeier. 2012. An operational application of the automatic feature extraction: The measurement of cracks in concrete structures. *The Photogrammetric Record* 17: 453–464.

D'Ayala, D. F. and S. Paganoni. 2011. Assessment and analysis of damage in L'Aquila historic city centre after 6th April 2009. *Bulletin of Earthquake Engineering* 9: 81–104.

Doan, H. T. X. and G. M. Foody. 2007. Increasing soft classification accuracy through the use of an ensemble of classifiers. *International Journal of Remote Sensing* 28: 4609–4623.

Dolce, M., A. Kappos, A. Masi, G. Penelis, and M. Vona. 2006. Vulnerability assessment and earthquake damage scenarios of the building stock of Potenza (southern Italy) using Italian and Greek methodologies. *Engineering Structures* 28 (3): 357–371.

Dupuis, J., S. Paulus, J. Behmann, L. Plumer, and H. Kuhlmann. 2014. A multi-resolution approach for an automated fusion of different low-cost 3D sensors. *Sensors* 14: 7563–7579.

Freitas, A. 2016. The use of infrared thermography and integrative tests for measuring rising damp. MSc diss., Universidade do Porto.

Furtado, A., C. Costa, A. Arêde, and H. Rodrigues. 2016. Geometric characterisation of Portuguese RC buildings with masonry infill walls. *European Journal of Environmental and Civil Engineering* 20 (4): 396–411.

Gatto, D. A., M. R. F. Goncalves, B. D. Mattos, L. Calegari, and D. M. Stangerlin. 2012. Estimativa da deterioração da madeira de assoalho de prédio histórico. *CERNE* 18: 4.

Gautam, D., H. Rodrigues, K. K. Bhetwal, P. Neupane, and Y. Sanada. 2016. Common structural and construction deficiencies of Nepalese buildings. *Innovative Infrastructure Solutions* 1 (1): 1.

Gonçalves, L., C. Fonte, E. Júlio, and M. Caetano. 2009. Assessment of the state of conservation of buildings through roof mapping using very high spatial resolution images. *Construction and Building Materials* 23: 2795–2802.

Gonzalez, R. and R. Woods. 2002. *Digital Image Processing*. Prentice Hall.

Gregorczyk, P. and P. B. Lourenço. 2000. A review on flat-jack testing. *Engenharia Civil UM* 9: 39–50.

Grinzato, E. 2012. IR thermography applied to the cultural heritage conservation. In *Proceedings of 18th World Conference on Nondestructive Testing*, 16–20 April, Durban, South Africa.

Grinzato, E., P. G. Bison, and S. Marinetti. 2002a. Monitoring of ancient buildings by the thermal method. *Journal of Cultural Heritage* 3: 21–29.

Grinzato, E., C. Bressan, S. Marinetti, P. G. Bison, and C. Bonacina. 2002b. Monitoring of the Scrovegni Chapel by IR thermography: Giotto at infrared. *Infrared Physics & Technology* 43 (3–5): 165–169.

Grossi, D. and E. Lama. 2015. Ultrasound technique to assess the physical conditions of the monument to Ramos de Azevedo, city of São Paulo, Brazil. *Revista Escola de Minas, Ouro Preto* 68 (2): 171–176.

Guidi, G., F. Remondino, M. Russo, F. Menna, and A. Rizzi. 2008. 3D modeling of large and complex site using multi-sensor integration and multi-resolution data. In *Proceedings of the 9th International Symposium on Virtual Reality, Archaeology and Cultural Heritage*, 2–5 December, Braga, Portugal.

Hall, D. L. and J. Llinas. 1997. An introduction to multisensor data fusion. *Proceedings of the IEEE*, 85: 6–23.

Kahraman, S., U. Ulker, and M. S. Delibalta. 2007. A quality classification of building stones from P-wave velocity and its application to stone cutting with gang saws. *Journal of the Southern African Institute of Mining and Metallurgy* 107 (7): 427–430.

Kandemir-Yucel, A., A. Tavukcuoglu, and E. N. Caner-Saltik. 2007. In situ assessment of structural timber elements of a historic building by infrared thermography and ultrasonic velocity. *Infrared Physics & Technology* 49: 243–248.

Köhler, W., A. Rohatsch, and J. Weber. 1996. The stone material of the west portal of St. Stephen's cathedral in Vienna: Petrological and ultrasonic investigations. In *Proceedings of the 8th International Congress on the Deterioration and Conservation of Stone*, ed. J. Riederer, Möller Druck und Verlag 2: 759–767.

Kordatos, E. Z., D. A. Exarchos, C. Stavrakos, A. Moropoulou, and T. E. Matikas. 2013. Infrared thermographic inspection of murals and characterization of degradation in historic monuments. *Construction and Building Materials* 48: 1261–1265.

Leckebusch, J. 2000. Two- and three-dimensional ground-penetrating radar surveys across a medieval choir: A case study in archaeology. *Archaeological Prospection* 7: 189–200.

Lerma, J. 2005. Automatic plotting of architectural facades with multispectral images. *Journal of Surveying Engineering* 31: 73–77.

Lerma, J. L., S. Navarro, M. Cabrelles, and V. Villaverde. 2010. Terrestrial laser scanning and close range photogrammetry for 3D archaeological documentation: The Upper Palaeolithic Cave of Parpalló as a case study. *Journal of Archaeological Science* 37: 499–507.

Li, Z., J. Chen, and E. Baltsavias. 2008. *Advances in Photogrammetry, Remote Sensing and Spatial Information Sciences:* 2008 *ISPRS Congress Book*. CRC Press.

Lillesand, T. M. and R. W. Kiefer. 2000. *Remote Sensing and Image Interpretation*. John Wiley & Sons.

Liñán, C. R., M. J. M. Conde, P. R. Hita, and F. P. Gálvez. 2015. Application of nondestructive techniques in the inspection of wooden structures of protected buildings: The case of Nuestra Señora de los Dolores Church (Isla Cristina, Huelva). *International Journal of Architectural Heritage* 9 (3): 324–340.

Liñán, C. R., and P. R. Hita. 1995. Evaluación del estado de la madera em obras de rehabilitación mediante técnicas de ultrasonidos y obtención de parâmetros resistentes. *Informes de la Construcción* 47 (440): 5–22.

Liñán, C. R., P. R. Hita, J. C. G. Cózar, and F. P. Gálvez. 2004. Diagnóstico mediante técnicas de ultrasonidos del forjado de madera del refectorio del Convento de Santa Clara en Carmona (Sevilla). *Informes de la Construcción* 55 (490): 17–28.

Lourenço, P. B., E. Luso, and M. G. Almeida. 2006. Defects and moisture problems in buildings from historical city centres: A case study in Portugal. *Building and Environment* 41 (2): 223–234.

Meola, C. 2013. Infrared Thermography in the Architectural Field. The Scientific World Journal, Volume 2013, Article ID 323948, 8 pages. doi: http://dx.doi.org/10.1155/2013/323948

Meola, C., R. di Maio, N. Roberti, and G. M. Carlomagno. 2005. Application of infrared thermography and geophysical methods for defect detection in architectural structures. *Engineering Failure Analysis* 12 (6): 875–892.

Merla, A., L. Di Donato, G. Martino, M. L. Rainone, and P. Signanini. 2009. An application of digital infrared thermography on historical cultural heritage damaged in the L'Aquila earthquake. In *Proceedings of 4th International Congress on Science and Technology for the Safeguard of Cultural Heritage in the Mediterranean Basin*, Cairo, Egypt.

Moreira, A. 2013. Estudo da Pedra Aplicada na Igreja da Senhora da Conceição (Porto). MSc diss., Universidade do Porto, Portugal.

Moropoulou, A., K. C. Labropoulos, E. T. Delegou, M. Karoglou, and A. Bakolas. 2013. Nondestructive techniques as a tool for the protection of built cultural heritage. *Construction and Building Materials* 48: 1222–1239.

Nebbia, M., A. Leone, R. Bockmann, M. Hddad, H. Abdouli, A. Masoud, N. Elkendi, et al. 2016. Developing a collaborative strategy to manage and preserve cultural heritage during the Libyan conflict. The case of the Gebel Nāfusa. *Journal of Archaeological Method and Theory* 23: 971–988.

Pantoja, P. 2016. A termografia infravermelha na salvaguarda de monumentos culturais como método não destrutivo de diagnóstico. MSc diss., Universidade Federal do Pará.

Paulus, S., J. Behmann, A. K. Mahlein, and H. Kuhlmann. 2014. Low-cost 3D systems: Well suited tools for plant phenotyping. *Sensors* 14: 3001–3018.

Pfeifer, N. and C. Briese. 2007. Laser scanning: Principles and applications. Institute of Photogrammetry and Remote Sensing, Publication database of the Vienna University of Technology, Austria. http://publik.tuwien.ac.at/files/pub-geo_1951.pdf (accessed January 2017).

Pratesi, F., D. Tapete, C. Del Ventisette, and S. Moretti. 2016. Mapping interactions between geology, subsurface resource exploitation and urban development in transforming cities using InSAR persistent scatterers: Two decades of change in Florence, Italy. *Applied Geography* 77: 20–37.

Ranalli, D., M. Scozzafava, and M. Tallini. 2004. Ground penetrating radar investigations for the restoration of historic buildings: The case study of the Collemaggio Basilica (L'Aquila, Italy). *Journal of Cultural Heritage* 5: 91–99.

RILEM. 2004. *In-Situ Stress Tests based on the Flat-Jack*. Bagneux, France: RILEM.

Rocha, Manuel, Joao J. Baptista Lopes, and Jorge Neves da Silva. 1966. "A New Technique For Applying the Method of the Flat Jack In the Determination of Stresses Inside Rock Masses." 1st ISRM Congress, 25 September–1 October, Lisbon, Portugal

Romão, X., A. A. Costa, E. Paupério, H. Rodrigues, R. Vicente, H. Varum, and A. Costa. 2013. Field observations and interpretation of the structural performance of constructions after the 11 May 2011 Lorca earthquake. *Engineering Failure Analysis* 34: 670–692.

Rossi, P. 1982. Analysis of mechanical characteristics of brick masonry tested by means of nondestructive in situ test. In *Proceedings of the 6th International Brick and Block Masonry Conference*, Rome.

Santos, C., T. M. Ferreira, R. Vicente, and J. A. R. Mendes da Silva. 2013. Building typologies identification to support risk mitigation at the urban scale: Case study of the old city centre of Seixal, Portugal. *Journal of Cultural Heritage* 14 (6): 449–463.

Sfarra, S., E. Marcucci, D. Ambrosini, and D. Paoletti. 2016. Infrared exploration of the architectural heritage: From passive infrared thermography to hybrid infrared thermography (HIRT) approach. *Materiales de Construcción* 66 (323): e094.

Simões, A., R. Bento, A. Gago, and M. Lopes. 2015. Mechanical characterization of masonry walls with flat-jack tests. *Experimental Techniques* 40 (3): 1163–1178.

Steel, A. M., M. J. Barnsley, and S. L. Barr. 2003. Inferring urban land use through analysis of the spatial composition of buildings identified in LiDAR and multispectral image data. In *Remotely-Sensed Cities*, ed. T. V. Mesev. Taylor & Francis.

Tapete, D. and F. Cigna. 2012. Rapid mapping and deformation analysis over cultural heritage and rural sites based on persistent scatterer interferometry. *International Journal of Geophysics* 19. doi:10.1155/2012/618609.

Tavukçuoğlu, A., A. Düzgüneş, E. N. Caner-Saltık, and Ş. Demirci. 2005. Use of IR thermography for the assessment of surface-water drainage problems in a historical building, Ağzıkarahan (Aksaray), Turkey. *NDT & E International* 38 (5): 402–410.

Tavukçuoğlu, A., S. Akevren, and E. Grinzato. 2010. In situ examination of structural cracks at historic masonry structures by quantitative infrared thermography and ultrasonic testing. *Journal of Modern Optics* 57 (18): 1779–1789.

Theodoridou, I., A. M. Papadopoulos, and M. Hegger. 2011. Statistical analysis of the Greek residential building stock. *Energy and Buildings* 43 (9): 2422–2428.

Valença, J., L. M. S. Gonçalves, and E. Júlio. 2013. Damage assessment on concrete surfaces using multi-spectral image analysis. *Construction and Building Materials* 40: 971–981.

Vicente, R., T. M. Ferreira, and J. A. R. Mendes da Silva. 2015a. Supporting urban regeneration and building refurbishment. Strategies for building appraisal and inspection of old building stock in city centres. *Journal of Cultural Heritage* 16 (1): 1–14.

Vicente, R., T. M. Ferreira, J. A. R. Mendes da Silva, and H. Varum. 2015b. In situ flat-jack testing of traditional masonry walls: Case study of the old city center of Coimbra, Portugal. *International Journal of Architectural Heritage* 9: 794–810.

Violla, S. 2012. Knowledge based protection for ancient city centres. 7th International Forum on Knowledge Asset Dynamics (IFKAD) / 5th Knowledge Cities World Summit (KCWS), Matera, Italy.

Wehr, A. and U. Lohr. 1999. Airborne laser scanning—An introduction and overview. *ISPRS Journal of Photogrammetry & Remote Sensing* 54: 68–82.

Więcek, B. and M. Poksińska. 2007. Passive and active thermography application for architectural monuments. In *Proceedings of the 8th Conference on Quantitative Infrared Thermography*, Padova, Italy.

Zheng, Y. 2015. Digital technology in the protection of cultural heritage Bao Fan temple mural digital mapping survey. In *International Archives of the Photogrammetry, Remote Sensing and Spatial Sciences*, Volume XL-5/W7, 25th International CIPA Symposium.

2 Characterisation of Building Stock and Its Pathologies

Case Study of the Historical City Centre of Leiria, Portugal

Tiago Gomes, Florindo Gaspar and Hugo Rodrigues

CONTENTS

2.1 INTRODUCTION

The study of existing buildings is of great interest nowadays due to the importance given to the preservation of built heritage.

The rehabilitation of buildings started a long time ago, but it was pioneered in monuments. From the twentieth century onwards, this concept was expanded, and it now encompasses urban built heritage, thus ensuring the identity of cities as well as their socio-economic, cultural and social values, passing on the history of several generations to future generations (Pereira 2013).

It is essential to safeguard the architectural heritage. For this purpose, it is necessary to develop a method, strategy and planning. A city is an urban network, composed of old buildings, covering a historical, ideological, architectural, artistic and cultural past. Since urban centres are the material testimony of the history of each city, when they undergo conservation, restoration or rehabilitation, their identity and original authenticity must be assured (Vicente et al. 2011).

In Portugal, since the 1970s, rehabilitation has been carried out in urban centres, with several successful cases, including Porto, Lisbon, Guimarães and Évora (Ramos and Lourenço 2004). However, existing urban rehabilitations are minimal compared with the number of degraded urban areas (Vicente 2008). Previous experiences of renovation or rehabilitation in other historic centres should serve as a basis for defining the methodologies and strategies to be used. However, each historic centre is a new case with numerous singularities, forcing a constant flexibility and capacity of adaptation (Lourenço et al. 2006).

Due to the lack of characterisation of the historical urban centre of Leiria, the work described in this chapter was developed with the purpose of characterising its historical city centre.

2.2 REHABILITATION AND REFURBISHMENT OF OLD CITY CENTRES

Construction-related interventions take place in very ambitious contexts in terms of cost constraints and deadlines (Indirli 2009). To guarantee all the objectives, it is necessary to plan, contextualise and conceive in detail the objectives that are intended to be carried out; to mobilising competent actors successfully; and to define leadership at the outset, with experience and compatible means, capable of ensuring and taking responsibility at all times for achieving the objectives (Appleton 2003). Several interventions in historical centres in Europe have been developed in the last years. In the following sections, two different experiences are briefly commented on.

The Historical Centre of Guimarães in Portugal is a reference, which has been classified as a UNESCO World Heritage Site. The main strategic lines of the intervention in the historical centre of Guimarães were the rehabilitation of public spaces and municipal buildings, giving them new form and new functions. The technical and financial support came from private initiatives, thus sustaining the two objectives that guide the intervention (Vicente 2008). (1) The rehabilitation of the historical centre of Guimarães aims to recover and preserve the built heritage of formal and functional quality while maintaining authenticity (using traditional materials and techniques). The historical nucleus was also adapted to the current way of life and its strict requirements, preserving its identity to maintain the architecture and the resident population (recovery and improvement of the dwellings – investing in rehabilitating and requalifying public spaces, equipment and infrastructure, with the strategy of attracting private interest). (2) The second objective was to maintain the entire resident population with better housing conditions.

The interventions in private buildings were carried out without excessive renovation. This also raised the possibility of a volumetric increase of existing buildings, making it economically and architecturally in keeping with what existed. The in situ learning allowed the development of skilled workmanship related to the traditional technologies. The operational strategy of this intervention had three lines of approach: (1) the control, stimulus and technical support of the particular initiatives; (2) conservation, restoration and modelling the rehabilitation of heritage and buildings of municipal responsibility; and (3) the requalification of public spaces.

The second experience is related to Berlin in Germany, where in the past, preference was given to urban renewal as an intervention strategy, which consisted of replacing the most affected and degraded parts of cities with new urban solutions, leading to the extinction of the memories of old German cities. Today, the city is quite different, being one of the most successful rehabilitation experiments. New key ideas emerged for action in the city, promoting (1) new buildings to fill the existing gaps in the city (voids left by the war and major renovation operations launched afterwards); (2) the rehabilitation of very degraded neighbourhoods through the recovery of old urban sectors and their buildings.

After these experiences a synthesis document was created, 'The 12 Principles for Cautious Urban Renewal', which was the main ethical reference of the interventions. Their success was due to the commitment and participation of the population, leading to the creation of local associations of residents. Support organisations were set up and funded, providing information and advice to residents interested in the rehabilitation process (on legal rights of residents and private rehabilitation initiatives, among others).

2.3 SURVEY APPRAISAL AND INSPECTION

The urban centres require knowledge that can support the rehabilitation of building stock, mainly because these buildings usually have to comply with specific requirements for an appropriate inclusion in the Historical City Centre. The global survey of the main architectural and structural characteristics, and the consequent elaboration of a diagnosis, can represent an important tool, giving a global image of the existent building stock. This preliminary work, although often discarded, is a fundamental tool for justifying the level of interventions to be carried out in both constructive and economic terms.

The International Council on Monuments and Sites (ICOMOS) (2001) and the ISO 13822 (2012) established guidelines and recommendations for assessment and preventative actions. Before carrying out any task, it is essential to know the characteristics of the existent buildings (structural behaviour, materials, value and identity), and it is necessary to gather historical information about the qualitative and quantitative nature of the techniques used in its construction, and the changes and conservation actions throughout its life, for a correct approach to the diagnosis of the interpretation of anomalies. The main concerns to be considered in a rational approach to a process of survey and inspection of old buildings are:

- Any construction is unique, so the needs for diagnosis, inspection and testing differ from case to case. The inspection strategy and techniques must be adaptable to the particularities of each construction.
- The choice, inspection and diagnosis registration tools must be adapted to the nature of the building and the physical limitations and the resources available. That is, the careful choice of what to inspect, as well as its depth

and detail, must be consistent with the objectives and scale of the planned intervention.

- Inspection and registration actions should consider the main objectives of the project. Any intervention, be it maintenance, alteration, reinforcement or renewal, should seek the best possible use of the investment made in the inspection phase (use of results produced).
- The definition of the strategy/methodology is clearer if it is translated into objective questions; for example, adaptability, construction phases and changes, among others. The answers will always be the result of the possible interpretations of the available information, thus avoiding mere speculation.
- Multidisciplinarity is currently a widely accepted principle. The contribution of a team of technicians with different skills, knowledgeable in diverse techniques and experiences, is an added value.
- The diagnosis process for inspection and registration can reach high levels of complexity. The quantity and quality of the information, the number of technicians and tasks involved in the process, and the coexistence and dynamics of different interpretations should not disrupt the objectives that were initially established.

In the present work, the inspection checklists were used to detect defects and problems, but also to register construction features and typology (support structure configuration, roof tiling/coverings, etc.). The inspection checklists used are based on the checklists proposed by Vicente et al. (2008) for the process of urban rehabilitation and the evaluation of the vulnerability and seismic risk of the construction of the downtown centre of Coimbra and for Seixal, also in Portugal (Santos et al. 2013). One of the biggest advantages in using these checklists is the fact that they have already been tested and adjusted for the Portuguese traditional building typology, and minor changes were implemented for the application in the historic city centre of Leiria (see Figure 2.1).

The samples from the historic city centre of Leiria are essentially divided into two parts: (1) visual and rapid inspection and registration with a sample of 302 buildings; (2) a detailed inspection and registration with a sample of 76 buildings.

2.4 CONSTRUCTIVE CHARACTERISATION OF BUILDINGS OF THE HISTORICAL CITY CENTRE OF LEIRIA

The city of Leiria is located in a very privileged geographical location, next to the river, between Coimbra and Santarém, with a castle at the top of a hill, which was the central point for the city. The area used in the study comprises 302 buildings (see Figure 2.2).

One of the main aims of the present work was to perform a characterisation of the traditional buildings that comprise the historic city centre of Leiria regarding the most important constructive elements: masonry walls, floor structures and roof structures.

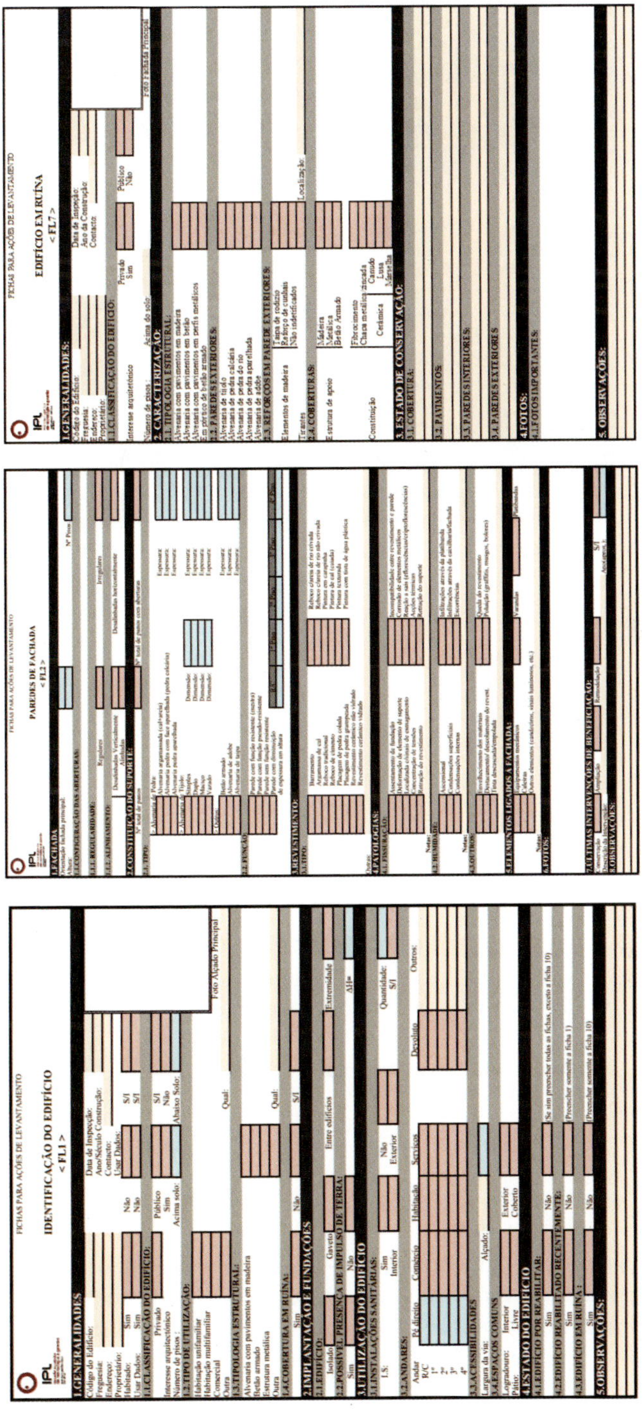

FIGURE 2.1 Examples of checklist.

FIGURE 2.2 Area under study.

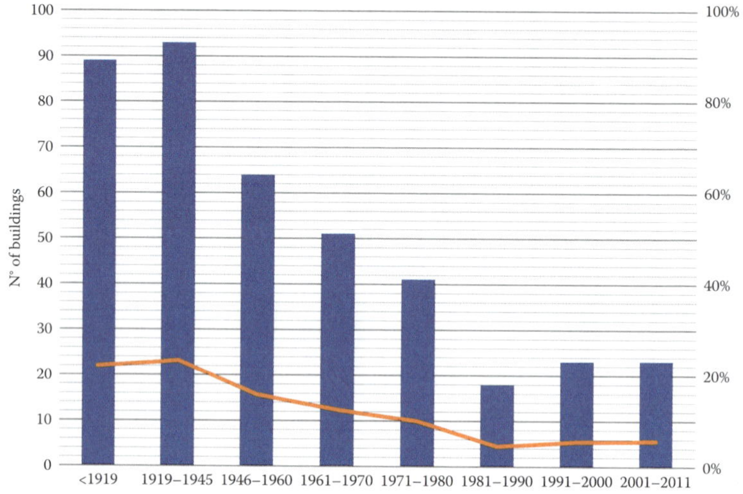

	<1919	1919–1945	1946–1960	1961–1970	1971–1980	1981–1990	1991–2000	2001–2011
N° of buildings	89	93	64	51	41	18	23	23
Percentage	22%	23%	16%	13%	10%	4%	6%	6%

FIGURE 2.3 Data of building construction in the historical city centre of Leiria.

Figure 2.3 shows the date of the construction for the buildings under study, whereby it is possible to verify that around 45% of constructions were developed before 1945. This information is very relevant to understanding the characteristics of different buildings and the need for the exhaustive study.

Regarding the classification of the building owners, only 2% (five buildings) are public buildings, leading to the conclusion that almost all of the buildings are private

FIGURE 2.4 Streets in the historic city centre of Leiria.

property (98% [297 buildings]). This also reveals the absence of public services in the city centre.

From the data collected, and regarding the type of building use, it is possible to verify that most of the buildings are multifamily with commercial activities (44%); the second category is only multifamily housing (34%). In general, the ground floor is used for services or commercial purposes; only the upper floors in most cases are intended as housing (see Figure 2.4). Regarding the number of storeys, the city centre is mainly composed of three-storey (63%) and two-storey (30%) buildings, with a few buildings that are one-storey (2%) and four-storey (5%). No original buildings were found with floors below ground; only one building with two floors below ground was observed.

2.4.1 Masonry Walls

Usually, the load-bearing walls of the buildings located in the historic city centre of Leiria are made up of very stiff stone masonry and with hydraulic mortar. Figure 2.5 represents examples of load-bearing walls that were found. The façade walls continue in general to be constituted by rubble stone, usually with wall thickness between 40 and 70 cm. Usually, the wall thickness decreases with height. Figure 2.6 presents the average thickness of the wall throughout the storeys. This is a common trend in old masonry buildings in old city centres (Vicente et al. 2015, Santos et al. 2013, Monteiro and Bento 2013).

The interior walls are usually composed of thinner walls incorporating embedded wooden structures, usually called *tabique* walls, made up of vertical boards or studs connected by laths through metal nails. This structure is then coated with an earth- or stone-based material. A lack of connections between interior and façade

FIGURE 2.5 Rubble masonry walls found in the city centre of Leiria.

walls was observed in many cases, which cannot ensure the strength and stability of the whole building.

Regarding the main sources of pathologies, it was possible to observe cracking of the masonry walls, in particular near the openings, mainly due to stress. Concentration of stress is the most common problem found in 80% of the buildings, and cracks related to settlement of the foundation were found in 54% of the buildings studied (see examples in Figure 2.7). Humidity and moisture are the

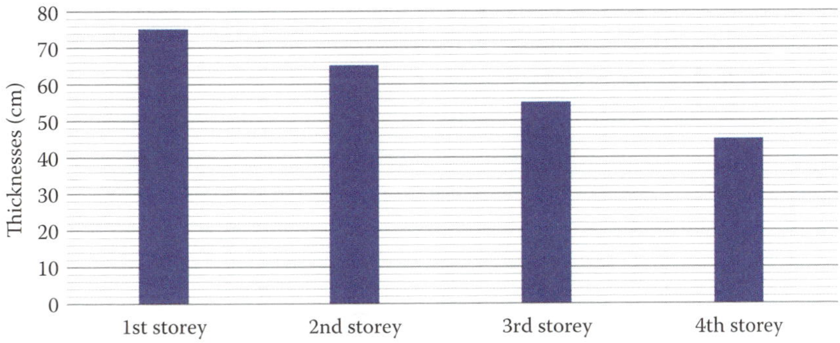

FIGURE 2.6 Wall thickness at different heights.

FIGURE 2.7 Degradation of the building façades.

cause of a different set of highly important pathologies. Based on the information collected, it was possible to observe that 67% of buildings had moisture problems due, among other causes, to malfunctions of the window frames, which were not able to guarantee their function; surface condensation was seen in 59% of buildings (see examples in Figure 2.8). These problems are in many cases related to the lack of maintenance of the buildings and also to improper use of the buildings. To sum up, it is possible to verify that cracking and humidity are the most common causes of pathologies observed in the existent buildings in the Leiria Historic City Centre. These facts point to the need for rehabilitation and refurbishment of the existent buildings.

FIGURE 2.8 Humidity in the façades of the buildings.

2.4.2 Floor Structures

The floors of the existent buildings are mainly composed of pine wood (see Figure 2.9). The floors directly support the resistant walls (façade walls), being composed of resistant wooden bars and nailed flooring. In around 30% of the analysed buildings, an air-box on the ground floor was found, which is a good practice, reducing the moisture that is transmitted through the ground and thus improving the conservation grade of the first floor and improving the comfort of the interior of the buildings.

The connection between the floors and the walls is a very important topic; there are many reports of anomalies, especially when in contact with humidity, or even due to excessive loads being applied to the floor. In many situations, they are not constructed in the best possible way, and in several cases they have only simple supports with a reduced embedded length.

FIGURE 2.9 Timber floors found in the city centre of Leiria.

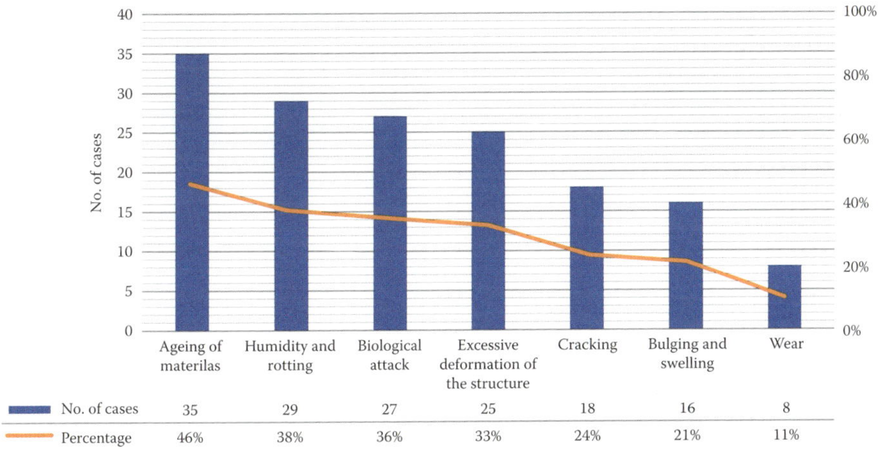

FIGURE 2.10 Common pathologies on timber floors.

In Figure 2.9, the anomalies found in timber floors are represented. The main factor is the ageing of the materials (46%), with more than 35% of the buildings having problems related to humidity and also biological attacks. These anomalies refer mainly to the finish/coating but also to the structure of the floors (see Figure 2.10).

Regarding the finishing of the timber floors, only anomalies in the ceramic tiles were found, related to the ageing of materials and their use and cracking. The problems observed in other materials (such as carpet and vinyl) are not relevant, and most of these elements are replaceable, since they are economical materials and easy to replace (see Figure 2.11).

FIGURE 2.11 Common examples of pathologies in the timber structures.

2.4.3 Roof Structures

The roof structures are the key element to keeping buildings properly preserved. In Leiria, the main cause of many of the buildings being in ruins or partially degraded is related to problems in the roofs, allowing water and other natural agents to penetrate the interior of the buildings, eventually degrading the resistant walls and the interior of the building. In the historic centre of Leiria, 89% of ruined buildings have pitched roofs, and only 11% are flat roofed. This is linked with the traditional use of ceramic tiles, with gable roofs in most of the cases and with two hips in around 90% of the cases.

One of the architectonical characteristics found in relation to the roofs is the eaves. These are typically constructed with a *canudo* tile, and the roof is composed of a similar tile type (in 38% of the cases) or of *marselha* or *lusa* tile. This is an

FIGURE 2.12 Different tiles used in the eaves and in the roof hips.

important aspect, because it gives an idea of the use of the same type in all the roofs (see Figure 2.12).

The majority of the structures of the roofs of the buildings of the historic centre of Leiria are composed of wooden elements (see Figure 2.13). Certainly, new and refurbished buildings have other materials that are not covered by the present study. Thus, 54% are supported with wooden beams, and 31% are composed of wooden trusses.

The ceramic coating of the roofs presents several anomalies. One of the most important problems found is related to broken tiles (30%), and a high percentage to

FIGURE 2.13 Different wooden supporting structures found in the city centre of Leiria.

misalignment of tiles (9%). Also, the excessive use of mortars in the ridge (45%) and in the eaves (37%) was found to be very common. This is a common practice to connect the tiles, but it does not allow ventilation between the tiles and the supporting structure, leading to a fast degradation.

However, looking at the wooden support structures, the main anomalies are related to the presence of moss and mould (80%), followed by the deformation of the support elements (44%) and finally, the degradation and ageing of the materials (32%). These pathologies usually lead to the failure of the waterproofing functional

FIGURE 2.13 (Continued) Different wooden supporting structures found in the city centre of Leiria.

requirement, and as mentioned before, this is fundamental for the conservation of the building and avoiding its degradation (see Figure 2.14).

2.5 MAIN CONCLUSIONS

The research work carried out has the objective of supporting potential investors in the area of conservation and/or rehabilitation of the historic centre of Leiria (the population, companies and the Leiria city council, among others).

Despite the absence of an integrated conservation policy, many of the old Portuguese urban centres constitute remarkable historical, archaeological, architectural, urban, social, scientific and technical sources of knowledge, making them a cultural heritage whose value transcends our country and the present generation, thus justifying the importance of safeguarding it. To ensure this, periodic conservation must be carried out, thereby avoiding serious degradation of historical centres. The smaller the rehabilitation solutions, the simpler they are and the greater their preservation, ensuring the original identity and authenticity.

To intervene in a historic centre, it is fundamental to elaborate a detailed strategic plan, adapted to the size of the study area, the objectives, and the intended detail of the actions and inspection tasks of the old buildings; only in this way is it possible to act correctly before large-scale work is required.

The characterisation of the built stock of the historic centre of Leiria through the registration of inspection records shows the importance of studying each historic centre individually, because although they have many similar values, there are other values with greater discrepancies, such as the materials used in the constructions, where the materials are local in many cases, as well as the construction techniques. For different materials and construction techniques, there are different anomalies that must be treated individually. Correct analysis of the symptoms and causes of a particular anomaly is important for an appropriate selection of solutions to eliminate

FIGURE 2.14 Pathologies found in ceramic tiles.

FIGURE 2.14 (Continued) Pathologies found in ceramic tiles.

the causes of this anomaly. Some of the most relevant differences among the studies are, for example, in façade wall anomalies. The present study shows a higher percentage due to factors such as pollution, ageing of materials, coating detachment or displacement and coating falling off, cracking and finally, humidity. In other case studies developed in Portugal, the humidity is a more relevant focus, being the main cause of pathologies. The study of Coimbra shows that there are many buildings with floors below the ground, while in Leiria, no buildings were found with this particularity. Among other differences, this underlines the importance of studying each historic centre.

And finally, with the present study, we could verify throughout the data obtained that the historic centre is in a high state of degradation and consequent abandonment, not in the commercial part but in the housing part, creating islands of activity in the historical centre without public services being available.

REFERENCES

Appleton, J. 2003. *Rehabilitation of Old Buildings: Pathologies and Techniques of Intervention*. Lisbon: Orion (in Portuguese).

ICOMOS. 2001. *Recommendations for the Analysis, Conservation and Structural Restoration of Architectural Heritage*. Paris International Scientific Committee for Analysis and Restoration of Structures of Architectural Heritage.

Indirli, M. 2009. Organization of a geographic information system (GIS) database on natural hazards and structural vulnerability for the historic center of San Giuliano Di Puglia (Italy) and the city of Valparaiso (Chile). *International Journal of Architectural Heritage* 3 (4): 276–315.

ISO13822. 2012. Bases for design of structures, Assessment of existing structures. Slovak Office of Standards, Metrology and Testing.

Lourenço, P. B., E. Luso, and M. G. Almeida. 2006. Defects and moisture problems in buildings from historical city centres: A case study in Portugal. *Building and Environment* 41 (2): 223–234.

Monteiro, M. and R. Bento. 2013. Seismic characterization and evaluation of an old masonry building. *SE-50EEE International Conference on Earthquake Engineering*, Skopje, Macedonia, 29–31 May 2013.

Pereira, A. 2013. Operações de Reabilitação de Edifícios Antigos: Organização de um Sistema de Informação Transversal a Todo o Processo (in Portuguese). Master's dissertation, Civil Engineering Department, University of Porto.

Ramos, L. F. and P. B. Lourenço. 2004. Modeling and vulnerability of historical city centers in seismic areas: A case study in Lisbon. *Engineering Structures* 26 (9): 1295–1310.

Santos, C., T. M. Ferreira, R. Vicente, and J. A. R. Mendes da Silva. 2013. Building typologies identification to support risk mitigation at the urban scale: Case study of the old city centre of Seixal, Portugal. *Journal of Cultural Heritage* 14 (6): 449–463.

Vicente, R., T. M. Ferreira, and J. A. R. Mendes da Silva. 2015. Supporting urban regeneration and building refurbishment. Strategies for building appraisal and inspection of old building stock in city centres. *Journal of Cultural Heritage* 16 (1): 1–14.

Vicente, R., H. Rodrigues, V. Humberto, and J. A. R. Mendes da Silva. 2011. Evaluation of strengthening techniques of traditional masonry buildings: Case study of a four-building aggregate. *Journal of Performance of Constructed Facilities* 25 (3). doi:10.1061/(ASCE)CF.1943-5509.0000164.

Vicente, R. S. 2008. Estratégias e metodologias para intervenções de reabilitação urbana: Avaliação da vulnerabilidade e do risco sísmico do edificadoda Baixa de Coimbra. PhD dissertation, Civil Engineering Department, University of Aveiro.

3 The Use of Multispectral Images and 3D-GIS Analysis for Monitoring the State of Conservation of Buildings

Case Study of the Historical City Centre of Leiria, Portugal

Luisa M. S. Gonçalves, Ana P. Falcão, Tiago Cerejo and Andreia Almeida

CONTENTS

3.1 INTRODUCTION

In the last century, rapid and uncontrolled urbanisation has had several consequences for the preservation of historical city centres, contributing to their degradation and to the separation of historic areas from the rest of the city instead of promoting a mutual physical integration and connectivity. In Bandarin and van Oers (2014), it is recommended that cities should be regarded as a legacy of historical layering and of cultural and natural values and attributes, from which opportunities can be generated whenever new urban development is conceived. Heritage preservation cannot be perceived as an obstruction to urban development, nor should 'the old' be an obstacle for 'the new' and an element to neglect, replace or destroy (Tapete, 2016).

In Portugal, due to urban sprawl and the construction of new modern residential areas, most of the historical downtown areas have been abandoned, presenting a high level of degradation. It is important to implement new strategies and policies to maintain the former character of cities and, at the same time, contribute to future resilience. For this reason, it is essential to know the state of conservation as well as the specific pathologies of each building element and its origin so as to act accordingly to ensure its functionality and durability and to have all this information accessible to decision makers and urban planners.

The traditional techniques used to obtain this information required an individual and in situ analysis for each building, making it time-consuming and demanding for large areas under assessment, such as roofs, which makes the inspection process even more difficult, since some parts of the roofs could be inaccessible. Nowadays, with increasing urban rehabilitation, it is important to explore other sources of information and develop new methods to identify the structural construction materials and nonstructural anomalies, which may be alternative or complementary to the traditional manual methods (Garcez, 2009; Gonçalves et al., 2009; Valença et al., 2014).

Several studies have shown that the use of multispectral images, obtained by photogrammetric processes or by satellite sensors, in detecting anomalies in the roofing materials and façades of buildings is rather promising (Hemmleb et al., 2006; Gonçalves et al., 2009; Lerma et al., 2011; Valença et al., 2013). In Gonçalves et al. (2009), high spatial multispectral images were used to get information about the conservation status of roofs. The method was tested in the classification of the historic city of Coimbra, in Portugal, which included more than 800 buildings, with very satisfactory results.

The management of architectural heritage information is also an issue that faces significant challenges, since architectural objects require an integrated representation of various types of information (historical, geographic and economic) to develop appropriate conservation strategies (Saygi et al., 2013). The existence of updated information, along with the use of technological platforms management and information visualisation, such as geographic information systems (GIS/3D-GIS), building information modelling (BIM) and city information modelling (CIM), decisively contributes to developing appropriate conservation strategies for historic city centres.

With the support of a case study, this chapter presents (1) the application of multispectral images of very large resolution, from different years, for the automatic detection of time changes of buildings' roofing materials and nonstructural anomalies using a hybrid methodology that is an evolution of the method developed by Gonçalves (2009); (2) the development of a 3D-GIS model that incorporates a geodatabase to store the information related to the history and the pathologies of each building of the model.

3.2 DATA SET AND CASE STUDY AREA

The study area is located in the historical Leiria city centre, with an area of approximately 0.40 km^2 and 671 buildings. The ancient downtown spreads from the slopes of the hill on which the castle is situated along the flat riverbanks and is occupied mainly by old buildings, some of which date from the sixteenth century to the nineteenth century, innumerable alleyways and some small squares.

To verify the type of the existing roofing materials and their state of degradation, multispectral aerial imagery with a spatial resolution of 0.5 m was used. The aerial images that were provided belong to the National Authority for Territorial Management (Direção Geral do Território) and were acquired with the four multispectral bands that were obtained in 2004 and in 2010, as shown in Figure 3.1.

The data used to produce the 3D-GIS city model, composed of 319 buildings, were (1) cartography information on a 1:10,000 scale; (2) architectural plans of the four historical buildings; (3) a database with the information of the nonstructural pathologies of building façades as well as the historical information; (4) a digital orthorectified aerial image with a spatial resolution of 0.5 m to be linked to the virtual 3D city model; and (5) façade images orthorectified with a spatial resolution of 0.012 m.

3.3 METHODS

The methodology proposed in this chapter is based on a multi-step procedure: (1) definition of the nomenclature (classes); (2) selection of the training sample set for the years 2004 and 2010; (3) image classification, based on a hybrid approach, to obtain the ceramic material (MBC) and anomalies (MBA) of the roofs of the buildings for the years 2004 and 2010; (4) evaluation of the classification accuracy; (5) comparison of MBC and MBA between 2004 and 2010, to assess the evolution of pathologies; (6) construction of the 3D-GIS model based on a developed geodatabase; and (7) classification of the building conservation status and mapping.

3.3.1 Multispectral Image Classification and Change Detection

A first image analysis was performed to identify the materials of the buildings' roofs and the presence of anomalies for the case-study area. The main roofing materials in the study area are dark ceramic roof tile, bright ceramic roof tile, fibrocement corrugated roof sheet, zinc roof panel, steel roof panel, concrete roof tile and roof shadow. Since the ceramic roof tile is the most representative roof material, the study focuses only on the identification of a ceramic roof tile with and without pathologies. This first analysis was important to define the nomenclature classes.

The subsequent step was the choice of the set of samples to train and evaluate the classifiers. A hundred samples (pixels) per class were selected: half were used for training and half for testing. Since this is a study to identify the changes occurring in buildings' roofing materials and pathologies between 2004 and 2010, it would be advisable to choose, in both images, samples with the same location. However, due to the different sun exposures affecting the aerial photographs, it was not possible to avoid considering samples in different locations for each year.

To evaluate the quality of the training set selected for each class, the spectral separability analysis was performed using scatter diagrams. This method allows, on the basis of the sample set, the recognition of classes with similar spectral signature. The quality of spectral separability between the classes is directly related to the quality of classification and decreases the ambiguity of the classifier in assigning a single

FIGURE 3.1 Aerial image (RGB 321) of the study area: (a) Typo-morphological groups of the historical centre of Leiria, overlaid on the image of 2010; (b) extract of the image of 2004; and (c) extract of the image of 2010.

class to the pixel. The choice of the training samples is an iterative process, assisted by spectral analysis. After obtaining satisfactory results, image classification was then carried out. The behaviour of the classifiers was analysed with the test samples.

The classification methodology developed involved two levels: the first, at pixel level, to extract the information of the types of roofing materials (RM) and roof anomalies (RA), and the second, at the level of the building (object), to obtain the map of the MBC and the MBA of the buildings' roofs. Since the idea was to assess the building changes that occurred between 2004 and 2010, four maps were obtained: the MBC and the MBA for the year 2004, and the MBM and the MBA for the year 2010.

The nomenclature used for the RM was dark ceramic roof tile (TC-D), bright ceramic roof tile (TC-B), fibrocement corrugated roof sheet (F), zinc roof panel (Z), steel roof panel (S), concrete roof tile (C) and roof shadow (S), and for the RA, ceramic roof tile with anomalies (TC-A), nondeteriorated ceramic roof tile (TC-ND) and ceramic roof tile with shadow (TC-S).

For the image classification, at the pixel level, a soft classifier was used, the Bayes' classifier (BAY). The process of extracting information using this type of data requires the classification of the images in a set of classes selected according to the characteristics and the resolution of the images and the purpose of the map. However, the application of different classifiers can provide different results. In the study of the anomalies of the façades performed by Valença et al. (2013), using this classifier, the authors obtained good results.

The classification at the building level to obtain the MBC and the MBA is carried out based on the arrangement of pixels inside the buildings and on the uncertainty of the pixel classification. For this classification, ancillary information in vector format (building polygons) was used, previously converted from vector to raster format. This building map in raster format was used as a mask to select only the information on the RM, the RA and their uncertainty inside the buildings. In summary, the MBC was obtained through the application of rules that incorporate the information on the membership grades assigned to the several classes at each pixel in the previous soft pixel-based classifications and the degree of uncertainty associated with these assignments.

The uncertainty information of the classification was obtained using an indicator of the classification uncertainty given by:

$$\mathrm{RI} = \frac{\max\limits_{i=1,\ldots,n}(p_i) - \dfrac{\sum_{i=1}^{n} p_i}{n}}{1 - \dfrac{1}{\eta}} \tag{3.1}$$

where

$(i = 1,\ldots,n)$ are the probabilities associated with the several classes

n is the number of classes under consideration

This indicator assumes values in the interval [0,1] and only depends on the maximum probability, the sum of all probabilities assigned to the class and the total number of classes. RI evaluates up to which point the classification is dispersed over more than one class and the degree of compatibility with the most probable class, providing information regarding the classifier's difficulty in assigning only one class to each pixel.

To obtain the MBC, the classes 'Buildings with Ceramic Roof Tile' (B-C) and 'Buildings without Ceramic Roof Tile' (B-NC) were used. The set of Rules 1 to 3 aimed to distinguish B-C from B-NC. To obtain the MBA, it was necessary to first acquire a map of MBC to ensure that only ceramic tile roofing with or without anomalies would be classified. The classes used to obtain the MBA were 'Buildings with Ceramic Roof Tile with Anomalies' (B-C-A) and 'Buildings with Ceramic Roof Tile Nondeteriorated' (B-C-ND), considering the latter to be in a good state of conservation. Rules 4 to 6 assign the B-C to one of two classes: B-C-A and B-C-ND. Table 3.1 presents the classification rules.

Finally, the analysis of the accuracy of the MBC and MBA maps produced was performed using a stratified random sample of 50 samples per class, considering the building as a unit sample.

To conclude, after getting the MBC and the MBA were obtained, the final maps obtained for the different dates were compared using geographic information system (GIS) tools, producing a map of changes. A map of the conservation status of the roofs was also produced, taking into consideration the percentage of pixels inside the buildings classified as ceramic roof tile with anomaly. The conservation status was classified as EC1 (the percentage of pixels as ceramic roof tile with anomaly is 0%), EC2 (the percentage of pixels classified as B-C-A is between 0% and 25%), EC3 (the percentage of pixels classified as B-C-A is between 25% and 50%), EC4 (the percentage of pixels classified as B-C-A is between 50% and 75%) or EC5 (the percentage of pixels classified as B-C-A is higher than 75%).

3.3.2 Construction of a 3D-GIS Model Geodatabase

The development of the 3D-GIS model started with the construction of the digital terrain model (DTM). For the whole urban area case study of approximately 0.40 km^2, a DTM with 1 m spatial resolution was produced using contour lines and height points. The DTM, built using ArcGIS tools, was then imported to the CityEngine (CE) software, and the digital orthorectified aerial image was projected on top of the DTM.

The next step was the 3D modelling of buildings, which were represented with diverse levels of detail (LOD). For all the 319 buildings of the historical city centre, geometry-models and differentiated roof structures and textures were created using CE software, and for the 110 buildings belonging to the heritage city centre, orthorectified images were projected onto the façades as textures to allow a more realistic representation (LOD 2). Architectural models and detailed indoors such as rooms, interior doors and stairs were only developed for four historical buildings and were made using the software BIM ArchiCAD. The highest level of detail (LOD 4) was necessary to create a model with information about its internal structure and the mutual relations between the architectonic elements, allowing a virtual visit to the

TABLE 3.1
Rules to Obtain MBC and MBA

No.	Rules	If True	If False
1	If more than 50% of the pixels of the RM, inside the buildings, are classified as 'Ceramic Roof Tile' with uncertainty less than 0.25.	Assign Class B-C	Apply Rule 2
2	If the percentage of pixels of the RM, inside the buildings, classified as 'Ceramic Roof Tile' with an uncertainty of less than 0.25, is higher than the percentage of pixels classified either as: 'Zinc Roof Panel'; 'Fibrocement Corrugated Roof Sheet'; 'Steel Roof Panel'; 'Concrete Roof Tile'; 'Roof Shadow'.	Assign Class B-C	Apply Rule 3
3	If the percentage of pixels of the RM, inside the buildings, classified as: 'Zinc Roof Panel'; 'Fibrocement Corrugated Roof Sheet'; 'Steel Roof Panel'; 'Concrete roof Tile' is less than 50% and if the percentage of pixels classified as 'Roof Shadow' is higher than 10% or if the percentage of pixels classified as 'Ceramic Roof Tile' is higher than 25%.	Assign Class B-C	Assign Class B-ND
4	If more than 50% of the pixels of the RA, inside the buildings classified as 'Ceramic Roof Tile' are 'Buildings with Ceramic Roof Tile with Anomalies' with uncertainty less than 0.25.	Assign Class B-C-A	Apply Rule 5
5	If the percentage of pixels of the RA classified as 'Ceramic Roof Tile with Anomaly', inside the buildings classified as 'Buildings with Ceramic Roof Tile' is higher than the percentage of 'Ceramic Roof Tile without Anomaly' and higher than the percentage of 'Roof Shadow'.	Assign Class B-C-A	Assign Class Rule 6
6	If the percentage of pixels of the RA classified as 'Ceramic roof Tile with Anomaly', inside the buildings classified as 'Buildings with Ceramic Roof Tile' is higher than the percentage of 'Ceramic Roof Tile without Anomaly' and the percentage of 'Roof Shadow' is higher than 50%.	Assign Class B-C-A	Assign Class B-C-ND

interiors of these buildings with access to historical information. The location of the buildings modelled with different levels of detail is presented in Figure 3.2.

The development of the virtual city model, implemented with the CE software, was done using 'Grammar-based rules/split grammar-base rules' or a 'procedural modelling' approach. The computer generated architecture (CGA) shape grammar of the CE is a programming language specified to generate architectural 3D content. The idea of grammar-based modelling is to define rules, or CGA rules within CE, that iteratively refine a design by creating more and more detail. The rules operate on shapes, which consist of geometry in a locally oriented bounding-box. The CE software provides an interface for shape grammar programming and commands such as 'extrude', 'split' or 'texture', allowing 3D complex architectural forms to be created.

Thus, the process included the digital terrain model (DTM) importation into CE software, the inclusion of the geodatabase with the buildings shapefile and attributes of the pathologies and historical data, the four buildings modelled in BIM ArchiCAD

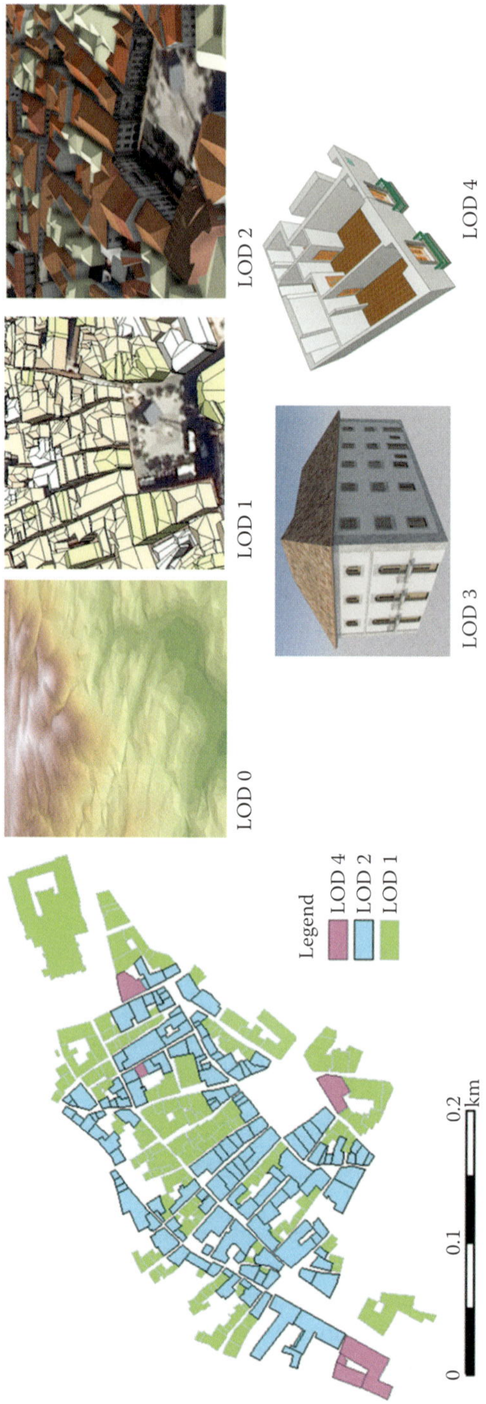

FIGURE 3.2 Spatial location of buildings devolved with different levels of detail.

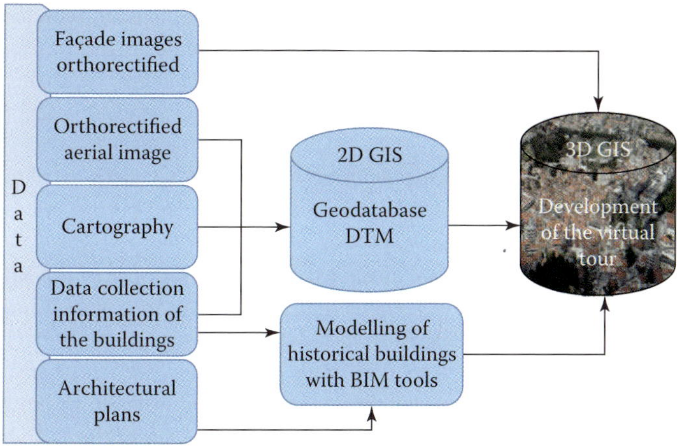

FIGURE 3.3 Methodology used to build the city model visualisation.

software and the façade orthophotos of the buildings. The geodatabase created to store the information related to the history and the pathologies of each building was subsequently integrated into the 3D model. All the attributes related to the historical and pathological information were combined with the 3D model. Figure 3.3 shows the implemented process.

3.4 RESULTS AND DISCUSSION

The accuracy evaluation of the MBC obtained with the stratified random samples set presented a global accuracy (GA) value of 93%, and with the selected samples, taking into account the degree of certainty, the GA was 72%.

When comparing the accuracy of the user and the producer for both sampling protocols (Figure 3.4), it is evidenced that the class of nonceramic roof tile (B-NC)

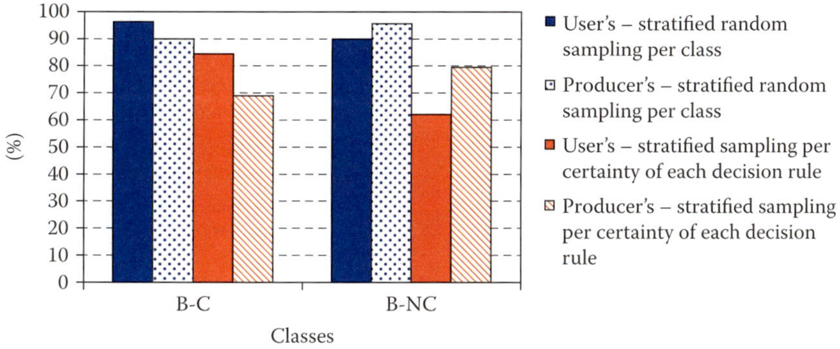

FIGURE 3.4 Graph showing the level of user's accuracy (UA) and producer's accuracy (PA) to assess the thematic accuracy of MBC, with stratified random samples and with samples taking into account the certainty of classification, for building with ceramic roof tiles (B-C) and building with nonceramic roof tiles (B-NC).

presents accuracy values for the producer higher than for the user, which means that, despite the correct identification of the nonceramic roofs existing in the study area and the reduced number of omission errors, there are ceramic roof tiles according to the reference data that have been misclassified as nonceramic tiles, originating commission errors.

The assessment of the thematic accuracy of the MBA with stratified random samples presented a GA value of 91%, while from samples, taking into account the degree of certainty, the GA value was 88%.

The comparison of the accuracy of the user and the producer of the MBA, shown in Figure 3.5, for both samples evidences that the B-C-A class of buildings presents higher values for producer accuracy than for the accuracy of the user. This means that, although the ceramic roofing tiles with anomalies were properly identified, there are rooftops without anomalies that were classified initially as having anomalies. The low values of accuracy of the producer and the user of the B-C-ND class of buildings are affected by the small size of the samples.

The results obtained with the multispectral image of the year 2004 show that 601 covers (90%) were identified as having ceramic tile coating, and for the image of the year 2010, 582 covers (87%) were identified, which corresponds to a decrease of 13%. Figure 3.6 shows the MBC for the years 2004 and 2010.

From the MBA information produced automatically, it was found that of building roofs with ceramic tiles for the year 2004, 537 (89%) were classified as ceramic tiles with anomalies, and 575 for the year 2010 (99%) (Figure 3.7). Comparing the anomalies in the building roofs also classified as ceramic tiles for the years 2004 and 2010, it was found that there were four rooftops classified as having anomalies in 2004, and in 2010 these roofs were classified as having no anomalies. This is due to incorrect classification or rehabilitation operations. The problems in the classification result from the high presence of shadow areas and different sun exposures in the multispectral image of 2004. In 2010, the existence of 58 roofs classified as having anomalies was verified, while in 2004 they were classified as having no anomalies,

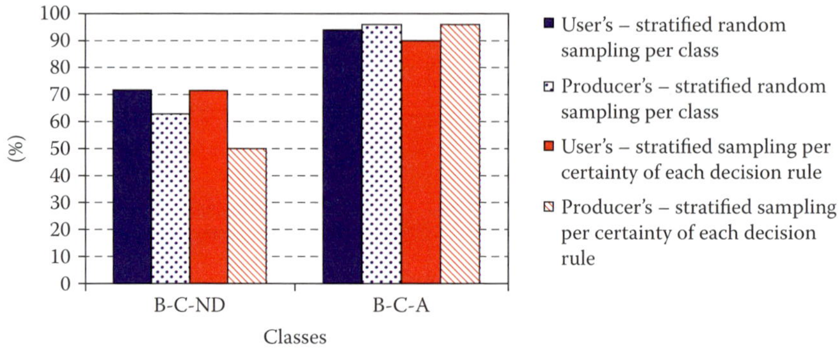

FIGURE 3.5 Graph showing the level of user's accuracy (UA) and producer's accuracy (PA) to assess the thematic accuracy of MBA with stratified random samples and with samples taking into account the certainty of classification of building ceramic roof tile nondeteriorated (B-C-ND) and with anomalies (B-C-A).

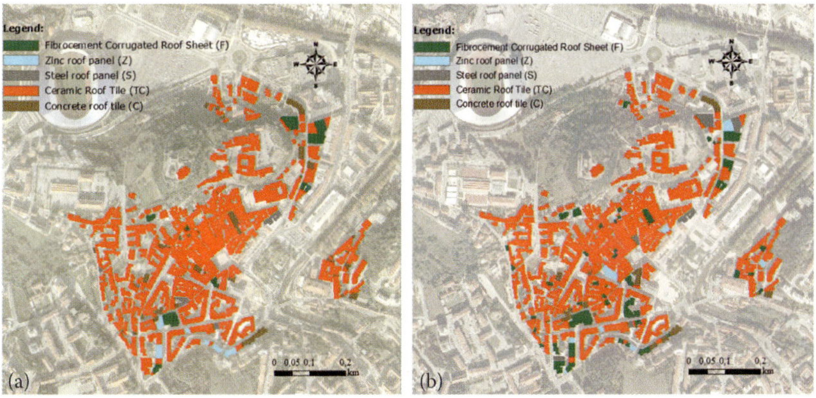

FIGURE 3.6 Maps of the types of roofing materials: (a) 2004 and (b) 2010.

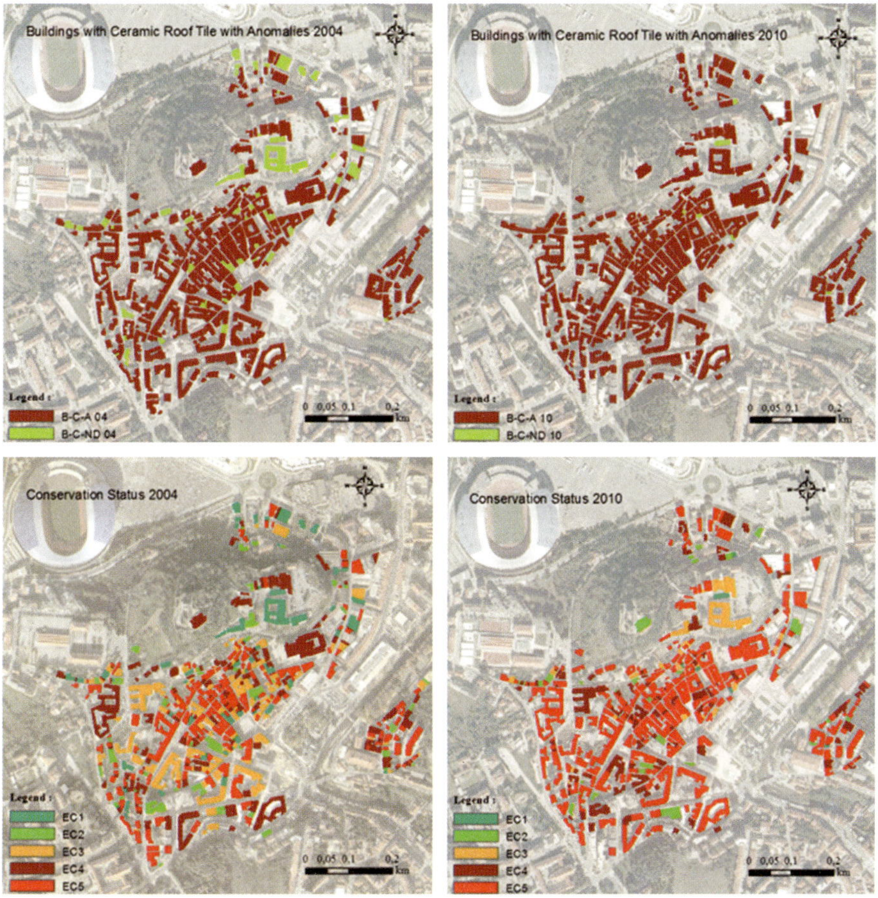

FIGURE 3.7 Map of ceramic anomalies (MBA) and map of conservation status of buildings' roofs.

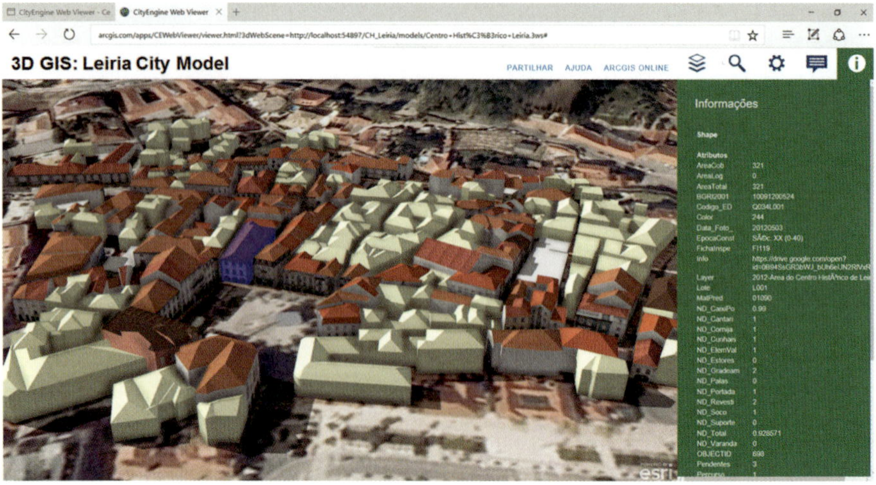

FIGURE 3.8 Example of information access throughout the 3D-GIS model.

which confirms the occurrence of anomalies over time and the scarcity of requalification activities in this study area.

The 3D-GIS city model of the historical city centre of Leiria was enriched with a collection of pathological information. The final model enables an interactive system for the management, integration and presentation of historical urban geoinformation. Figure 3.8 shows an example of a 3D building map in which it is possible to consult information such as the type of usage, period of construction, global level of degradation, and so on. The access to this kind of information could be used by municipal technicians to support their decisions related to the conservation of the historical city centre.

3.5 CONCLUSIONS

The results obtained show that it is possible to retrieve information about the existence of anomalies and different types of roofing materials by using multispectral images to extract ancillary information to assess the state of conservation of building roofs with high accuracy. The study proved that the classification decision rules developed by Gonçalves (2009), when applied to other case studies, continue to provide high values of accuracy.

The use of multispectral images from different years is advantageous, because it allows the assessment of the evolution of anomalies and changes occurring in building roofs, such as changes in the type of coating and the existence of rehabilitation activities.

The multispectral aerial images with 0.5 m spatial resolution allowed the detection of different types of roofing materials. However, this only enabled the identification of roofs with anomalies and did not allow the identification of the different types of pathology. In the future, the increasing technological innovation and the use of images with a higher spatial resolution, or hyperspectral images, may provide a way to distinguish the different types of roof anomalies.

The 3D-GIS city model of the historical city centre of Leiria was enriched with a collection of pathological information. The final model enables an interactive system for the management, integration and presentation of historical urban geoinformation.

The development of Web services for the model's contents could extend the ways in which the city model can be accessed by the municipality technicians and other partners contributing to the preservation and promotion of the built capital, heritage identity and memory.

ACKNOWLEDGEMENTS

This work has been supported by the Fundação para a Ciência e a Tecnologia (FCT) under project grant UID/MULTI/00308/2013.

REFERENCES

Bandarin, F. and R. van Oers. 2014. *Reconnecting the City: The Historic Urban Landscape Approach and the Future of Urban Heritage*. Oxford: John Wiley & Sons.

Garcez, N. 2009. Sistema de inspeção e diagnóstico de revestimentos exteriores de coberturas. Master's diss., Instituto Superior Técnico, Lisbon University.

Gonçalves, L. 2009. Integração da Incerteza na Classificação e Avaliação da Exactidão Temática de Imagens Multiespectrais: Aplicação à Avaliação do Estado de Conservação do Património Edificado da Baixa de Coimbra. PhD diss., FCTUC, Coimbra University (in Portuguese).

Gonçalves, L., C. Fonte, E. Júlio, and M. Caetano. 2009. Assessment of the state of conservation of buildings through roof mapping using very high spatial resolution images. *Construction and Building Materials* 23: 2795–2802.

Hemmleb, M., F. Weritz, A. Schiemenz, A. Grote, and C. Maierhofer. 2006. Multi-spectral data acquisition and processing techniques for damage detection on building surfaces. *Proceedings of ISPRS Commission V Symposium*, 25–27 September, Dresden, Germany.

Lerma, L., M. Cabrelles, and C. Portalés. 2011. Multitemporal thermal analysis to detect moisture on a building façade. *Construction and Building Materials* 25: 2190–2197.

Saygi, G., G. Agugiaro, M. Hamamcıoğlu-Turan, and F. Remondino. 2013. Evaluation of GIS and BIM roles for the information management of historical buildings. In *ISPRS Annals of the Photogrammetry, Remote Sensing and Spatial Information Sciences* Volume II-5/W1, XXIV International CIPA Symposium.

Tapete, D. 2016. Reconnecting the city. *Cities Spatial Issue* 56: 99–100.

Valença, J., D. Dias-da-Costa, L. M. S. Gonçalves, E. Júlio, and H. Araújo. 2014. Automatic concrete health monitoring: Assessment and monitoring of concrete surfaces. *Structure and Infrastructure Engineering: Maintenance, Management, Life-Cycle Design & Performance* 10 (12): 1547–1554.

Valença, J., L. M. S. Gonçalves, and E. Júlio. 2013. Damage assessment on concrete surfaces using multi-spectral image analysis. *Construction and Building Materials* 40: 971–981.

4 Mechanical Characterisation of Masonry Walls with Flat-Jack Tests

Patrick Pinheiro, Paulo Fernandes, Pedro Santos and Hugo Rodrigues

CONTENTS

4.1 INTRODUCTION

Flat-jack testing is considered to be a nonintrusive test technique (the opening of small and thick slits) that allows in situ determination of the installed stress and the deformability characteristics of masonry. At present, this test is often known as a test associated with masonry, but it actually had its origins in the mechanics of rocks, undergoing an adaptation by the researcher Paolo Rossi in the mid-1980s (Rocha et al. 1966, Rossi 1982).

This testing technique has proved to be reliable and effective and has been seen in several test campaigns around the world (Gregorczyk and Lourenço 2000). The level of confidence in the results is mainly due to reduced disturbances in the delimited area of the wall, where the mechanical properties are evaluated (Almeida et al. 2012). The test protocol for flat-jack testing is defined by the American Society for Testing Materials (ASTM 1991) and Réunion International des Laboratoires et des Matériaux, Systémes de Construction et Ouvrages (RILEM 2004).

Flat jacks can be considered a slightly destructive technique when used as a diagnosis tool with brick and stone masonry structures (Binda and Tiraboschi 1999). Initially, they have been applied in combination with other nondestructive tests for the assessment of structures with certain particularities, such as masonry bridges (Orbán and Gutermann 2009, Costa et al. 2016) or monuments (Binda et al. 2003, Carpinteri et al. 2009), among many other cases, and in combination with acoustic tests, ground-penetrating radar, and finite element method (FEM), among others (Ranalli et al. 2004, Carpinteri et al. 2009). The results of a testing programme show the efficiency of various nondestructive test methods, which has proved to be valuable information with reference to the condition of the elements under study and also helps to verify the basic input parameters for structural analysis and assessment (Bosiljkov et al. 2010).

Flat jacks have also been widely used to characterise rubble masonry walls in city centres around the world. For example, several studies have been found for different cities, such as Lisbon (Simões et al. 2012, 2016), Coimbra (Vicente et al. 2015), Seixal, Porto (Almeida et al. 2012), and Tentugal (Lourenço et al. 2005), all of which are located in Portugal. The review of the results obtained showed a significant difference in the results found, which led to the need to adapt testing in other cities with different materials and local construction techniques. As such, this chapter will show the applications of flat-jack testing in three buildings situated in the historic centre of Leiria.

4.2 EXPERIMENTAL CAMPAIGN

To assess the mechanical properties of the walls from representative masonry buildings located in the city centre of Leiria, three buildings were selected to perform the in situ flat-jack tests (four single and four double tests) (see Figure 4.1). Based on the information collected to characterise the typical construction used in the historic centre of Leiria, it was clear that the use of unreinforced rubble masonry walls and wood for the floors and roofs was most appropriate. The selection of the buildings followed the three criteria proposed by Vicente et al. (2015):

- The buildings had to have minimum conditions of structural stability that did not compromise the testing and safety of the workers involved.
- The walls tested were representative of the typical Historical Centre of Leiria (HCL) type of masonry.
- Only zones complying with the requirements of ASTM C1196-91 (ASTM 1991) and RILEM LUM.D.2 (RILEM 2004) were tested.

The flat jacks used are represented in Figure 4.2 and are made of one steel sheet folded in the middle and welded at the edges, creating an envelope bladder to be filled with oil. The pressure of the oil inside the flat-jack is measured and controlled by a manual pump. According to the ASTM (ASTM 1991) and RILEM (RILEM 2004) standards, the flat jacks have an inherent stiffness, which resists expansion when pressurised. Therefore, the flat-jack oil pressure is greater than the stress that is applied to the masonry wall (Simões et al. 2016). The average stress applied to the masonry (σ_m) is equal to the flat-jack pressure (p) multiplied

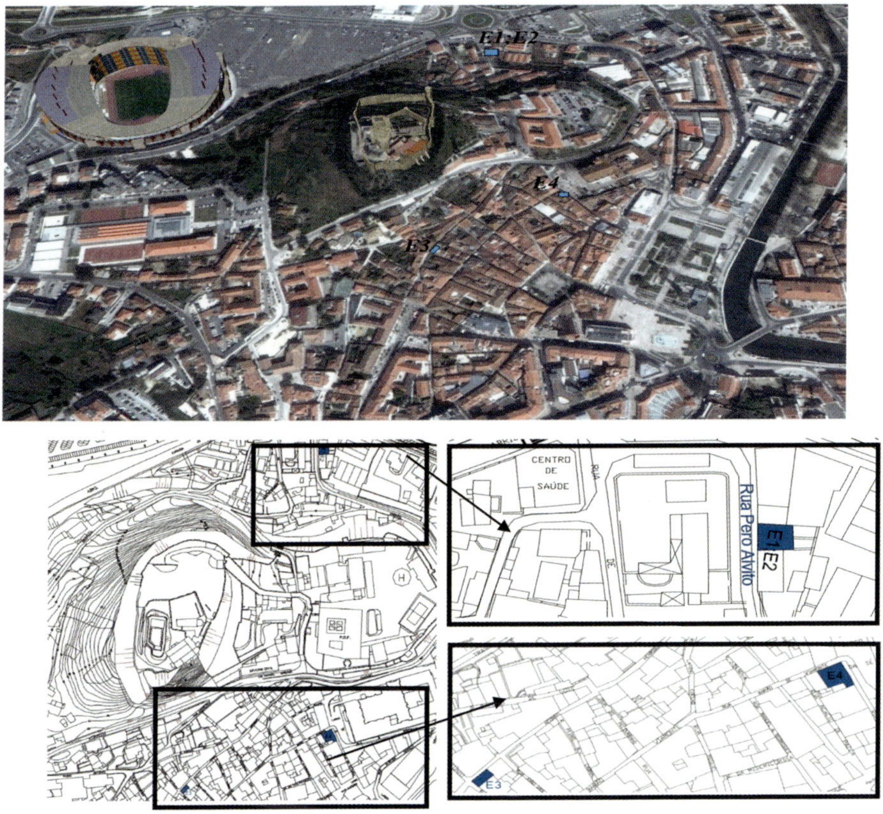

FIGURE 4.1 Location of the buildings tested.

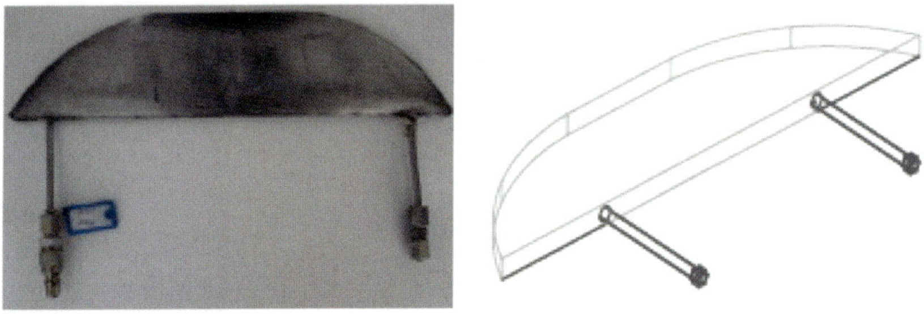

FIGURE 4.2 Flat-jacks configuration used in the present study.

by factors that account for the physical characteristics of the flat-jack; namely, an a-dimensional calibration factor (K_m) and the ratio K_a between the effective area of the flat-jack in contact with the masonry (A_{je}) and the area of the cut (A_{cut}), according to Equation 4.1.

$$\sigma_m = pxK_mxK_a \tag{4.1}$$

4.3 BRIEF DESCRIPTION OF THE SELECTED BUILDINGS

The Pero Alvito building (E1 and E2) dates from the construction period of the twentieth century (1940–1970) and is situated on the north-east slope of the historic centre of Leiria. It has a square plan configuration consisting of two floors and is in a reasonable state of conservation (Figure 4.3). Two different walls were subjected to flat-jack tests.

The D. Afonso Henriques building (E3) (Figure 4.4) and the E4 building (Largo da Sé) (Figure 4.5) are nineteenth-century buildings and are located in the central nucleus of the historic centre of Leiria with a rectangular plan configuration. They are constituted of three floors, and they were tested in a reasonable state of conservation. The roofs of buildings E3 and E4 are made up of a ceramic cover (*canudo* tile) and structural walls of rubble stone masonry, which are partitioned by tabique partition walls and with timber floors.

4.4 PRELIMINARY WORKS

The fieldwork, which was considered preliminary work, was developed for the flat-jack tests to be performed correctly. The first step was to choose the building and define which wall was going to be tested. Once the location was defined, the following steps were undertaken:

1. The coating layer was removed.
2. The limits of the inspection window to be opened were marked, and the coating layer was carefully removed to avoid causing damage to the wall (Figure 4.6a and b). Since it was also intended to characterise the percentage

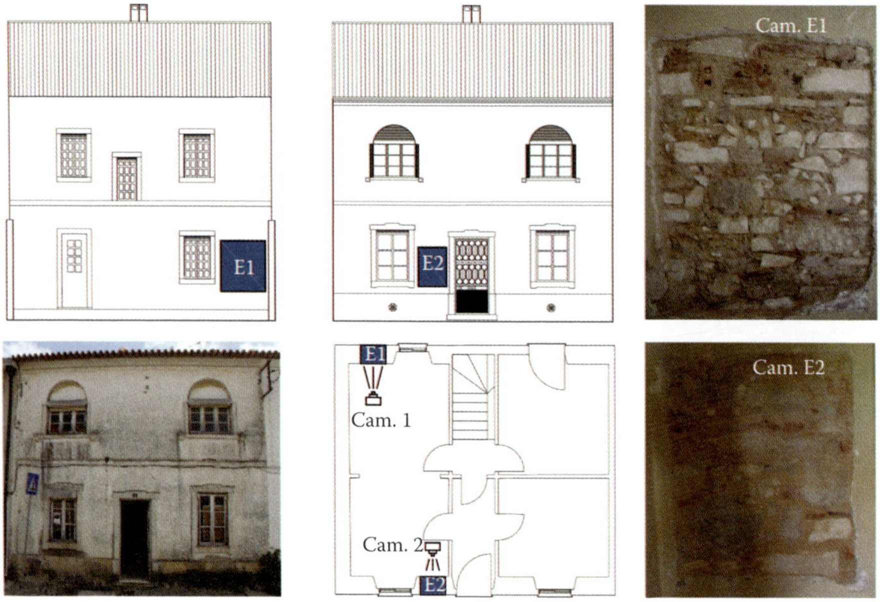

FIGURE 4.3 Masonry walls tested on the Pero Alvito building (E1 and E2).

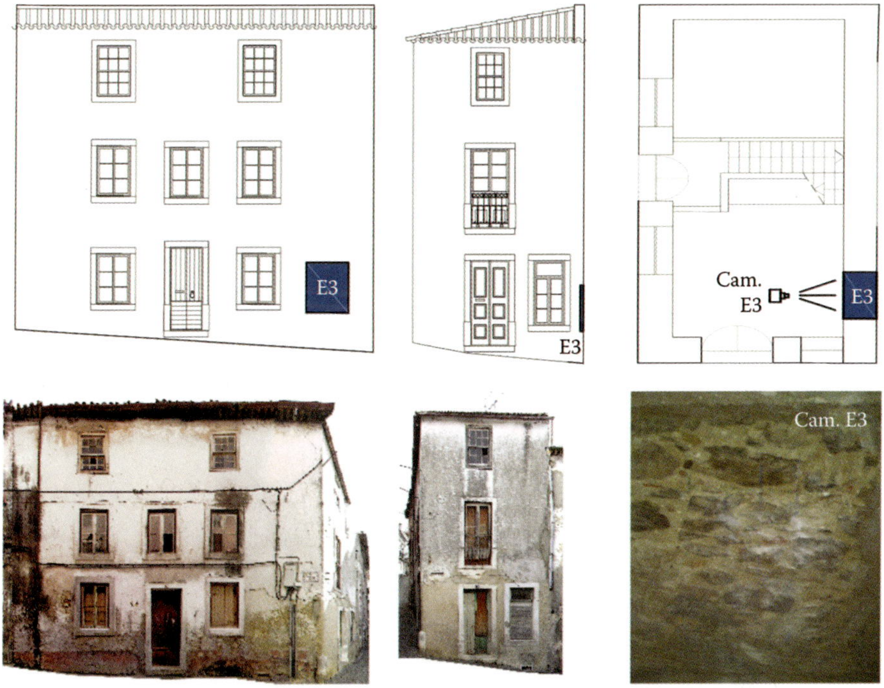

FIGURE 4.4 Masonry walls tested on the D. Afonso Henriques building (E3).

FIGURE 4.5 Masonry walls tested on the Largo da Sé building (E4).

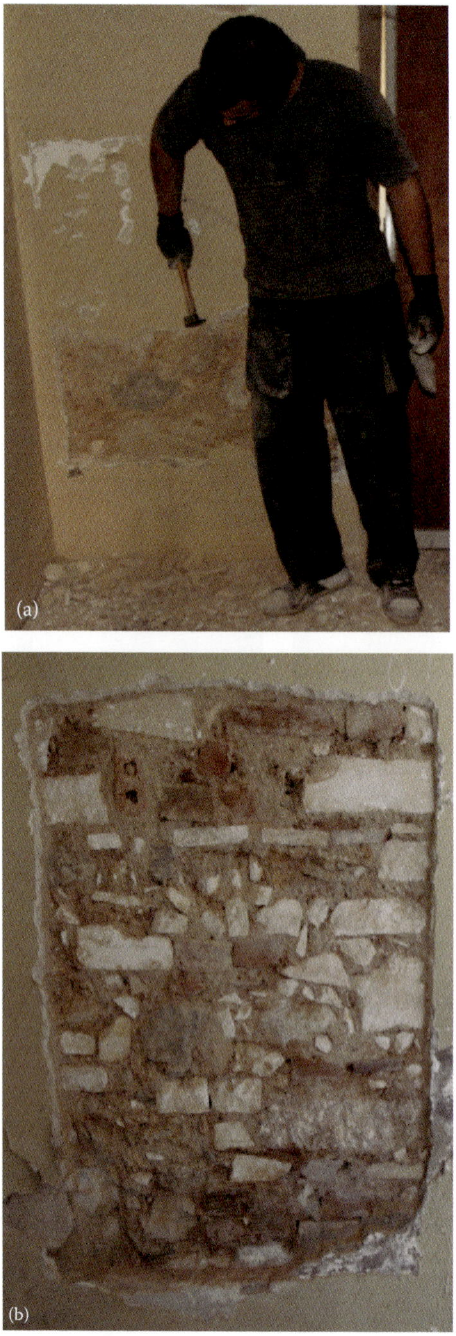

FIGURE 4.6 Removal of the coating layer: (a) during the removal of the coating layer; (b) final tests window.

of masonry constituents and to have a clear view of the wall during the tests, a brushing process was carried out until there was a clear perception of the masonry wall's constituents and limitations.

3. Alignment locations were set. First, the zone to perform the horizontal cuts in the wall was defined (Figure 4.7a). Once marked, three vertical alignments identified as A1, A2 and A3 and a horizontal alignment corresponding to the identification A4 (Figure 4.7b and c) were defined. The marking of the alignments allowed the marking of points to fix the rods, which were then used to support the instrumentation to perform the test.

4.5 SINGLE FLAT-JACK TEST

4.5.1 Introduction

After the wall to be tested was chosen and the alignments defined, the sensors were installed in the three vertical alignments, and everything was set for the simple test. The procedure is shown in Figure 4.8.

The next step was to make a cut in the masonry wall. This process proved to be the most difficult because of the weight of the motor-disc and also because the cutting position did not provide the best working position. Another aggravating factor was the dust released as a result of the execution of the cut, which, in addition to hindering the vision during the cutting process, also implied the need to protect all the testing equipment.

The flat jacks (with the chemicals for the contact area evaluation) were introduced into the cut. Once the flat-jack was put in place, the test was finally performed with increments of 0.05 MPa in each load cycle until the initial distances were established.

4.5.2 Stress Estimation

From the displacement results, the in situ stress is evaluated from the average of the three deformation measurements obtained from the alignments. The results in terms of evolution are presented in Figures 4.9 through 4.12 and summarised in Table 4.1.

It should be mentioned that the dimension of the stone element and the arrangement in the test were changed for the development of alternative load paths, which could compromise the reliability of the test results. There was also the fear of invalidating the use of one of the measurements obtained. This fact is also referred to in other studies in the literature (Vicente 2008).

Table 4.1 also presents the estimated stress based on the geometrical data compared with the stress levels observed in the single tests. The deviations are around 25%. This may be related to a large number of reasons that make it difficult to calculate the vertical stress level in simple assessments: the quality, heterogeneity and constitution of the masonry element under test, not forgetting the nonuniform destitution of the vertical loads, among many others.

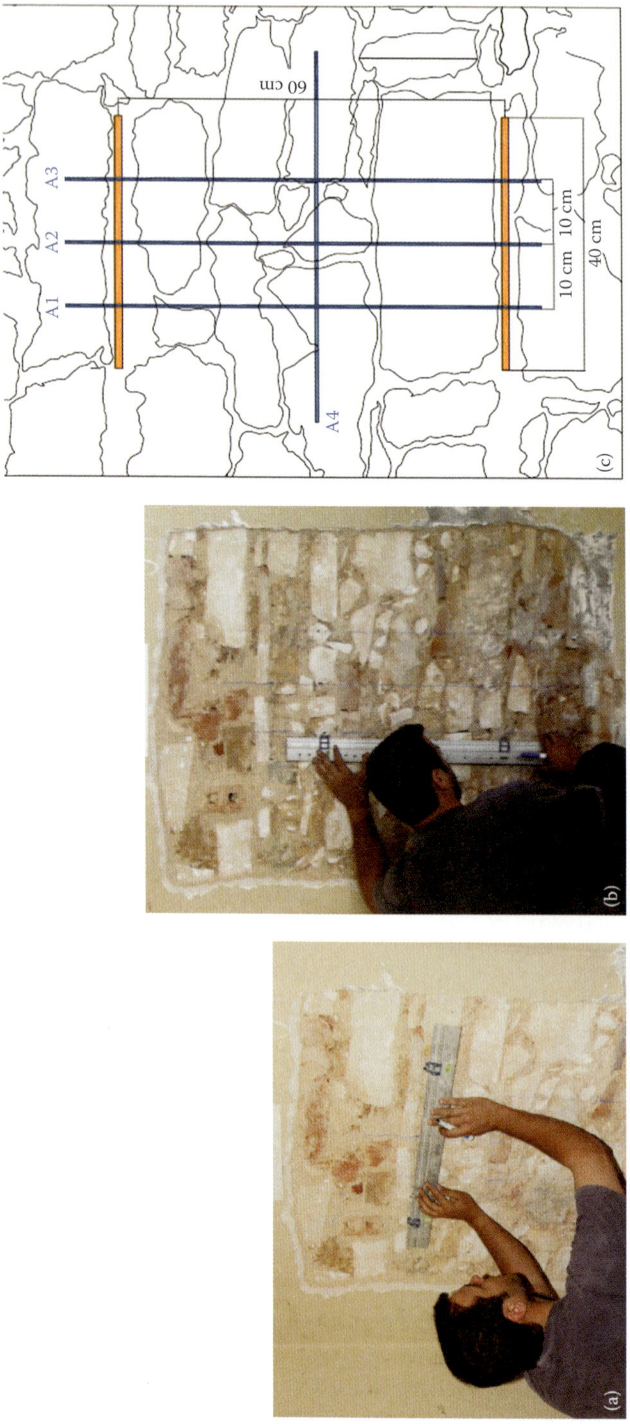

FIGURE 4.7 Identification cuts and alignments: (a) marking of cuts; (b) marking the alignments; and (c) identification of alignments and their distances.

FIGURE 4.8 Schematic protocol of simple flat-jack tests: (a) recording of initial distances; (b) opening the cut; (c) cleaning and removing the dust inside the cut; (d) final overview of the cut; (e) measurement of distances after opening the cut and placement of the flat-jack; and (f) reinstatement of the initial distances.

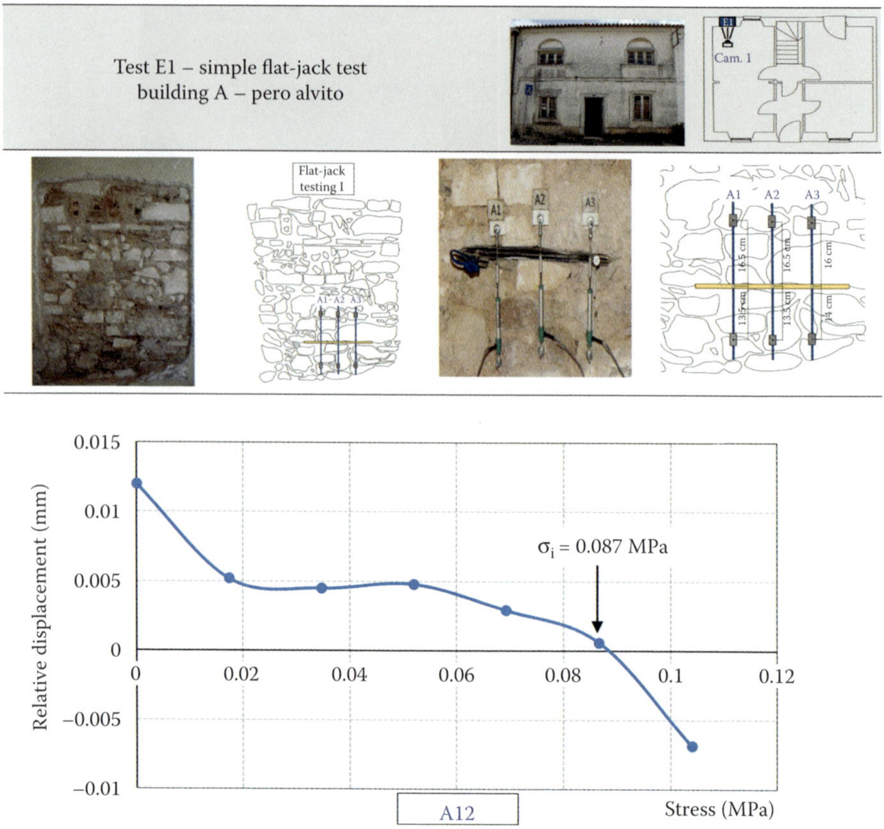

FIGURE 4.9 Single test E1: test representation and evolutions of displacements.

4.6 DOUBLE FLAT-JACK TEST

The execution of the double test is started after the simple test. This is done with another opening in the same wall that is being studied after a second cut (Figure 4.13) that is parallel to the first one, followed by the placement of the flat-jack (Figure 4.13a). The next step was the instrumentation of the double test according to ASTM C 1997-91 (ASTM 1991) by placing the three vertical potentiometers and one in the horizontal position. The flat jacks were connected to the hydraulic system, allowing increases of equal pressure value in both jacks. Three loading and unloading cycles of 0.1 MPa were then applied. The tests were finalised by the absence of reaction in the hydraulic pump. The overview of the double test is shown in Figure 4.13.

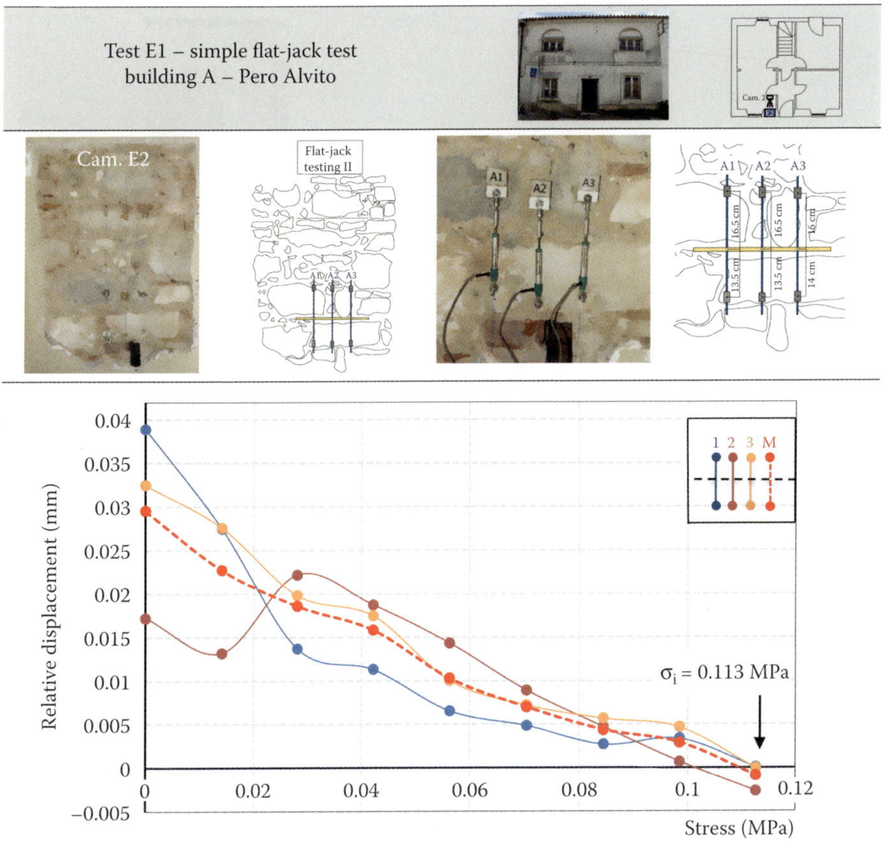

FIGURE 4.10 Single test E2: test representation and evolutions of displacements.

The results obtained allow the evaluation of the mechanical characteristics of the masonry under uniaxial loading and also the estimation of the elastic modulus and compressive strength.

Figures 4.14 through 4.17 present the average stress-strain results for tests E1 to E4, and it can be seen that the in situ stress estimated in the single test is generally much lower than the maximum stress attained for each test. Table 4.2 presents the ratio between the in situ stress estimated in the single test and the maximum strength, showing a safety factor higher than 5 in all cases.

The evaluation of the elasticity modulus of the masonry walls tested was obtained through graphical analysis, estimating the stress-strain curves obtained in the double tests. Figure 4.18 shows the different ways of evaluating the elasticity modulus: the

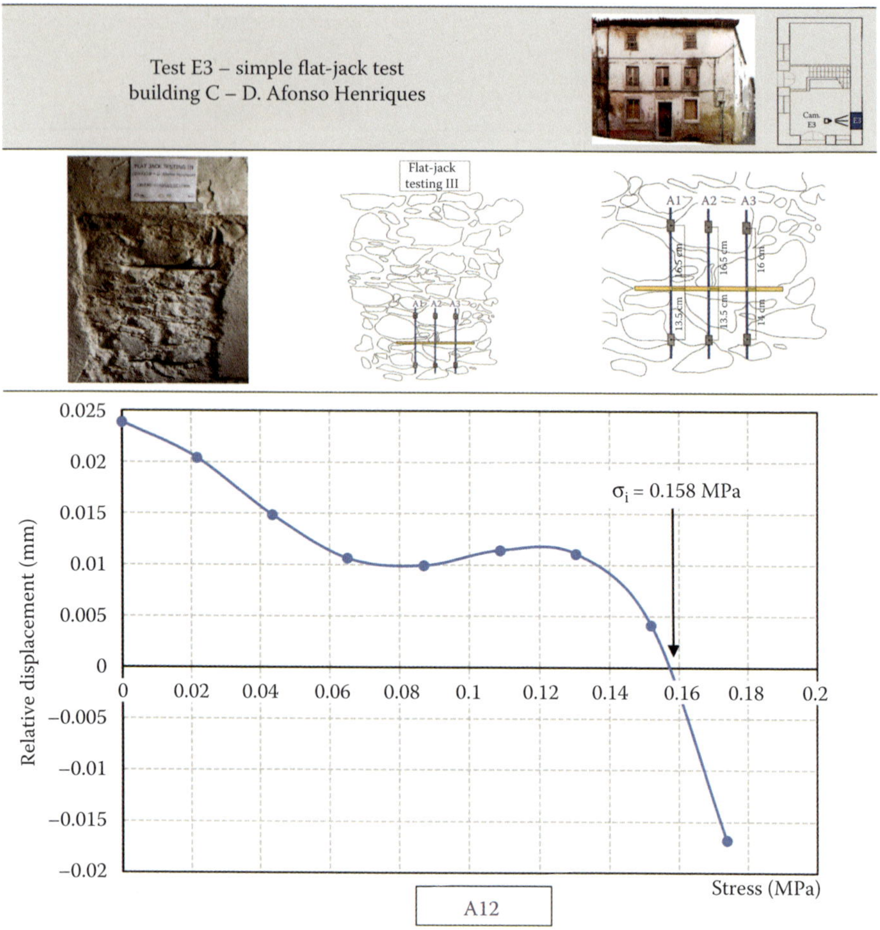

FIGURE 4.11 Single test E3: test representation and evolutions of displacements.

slope of the secant stiffness corresponding to different percentages of the maximum strength. The modulus of elasticity of more interest in this investigation was that corresponding to the secant stiffness between 30% and 60% of the maximum strength, because in this zone of the stress-strain, the curve presents linear behaviour. Table 4.3 shows the estimation of various modulus of elasticity values related to the stress-strain curves.

4.7 FINAL COMMENTS

An experimental campaign with flat jacks was carried out to assess the mechanical properties of old masonry buildings in the historical centre of Leiria. The campaign included tests on four rubble stone masonry walls from traditional buildings, and in

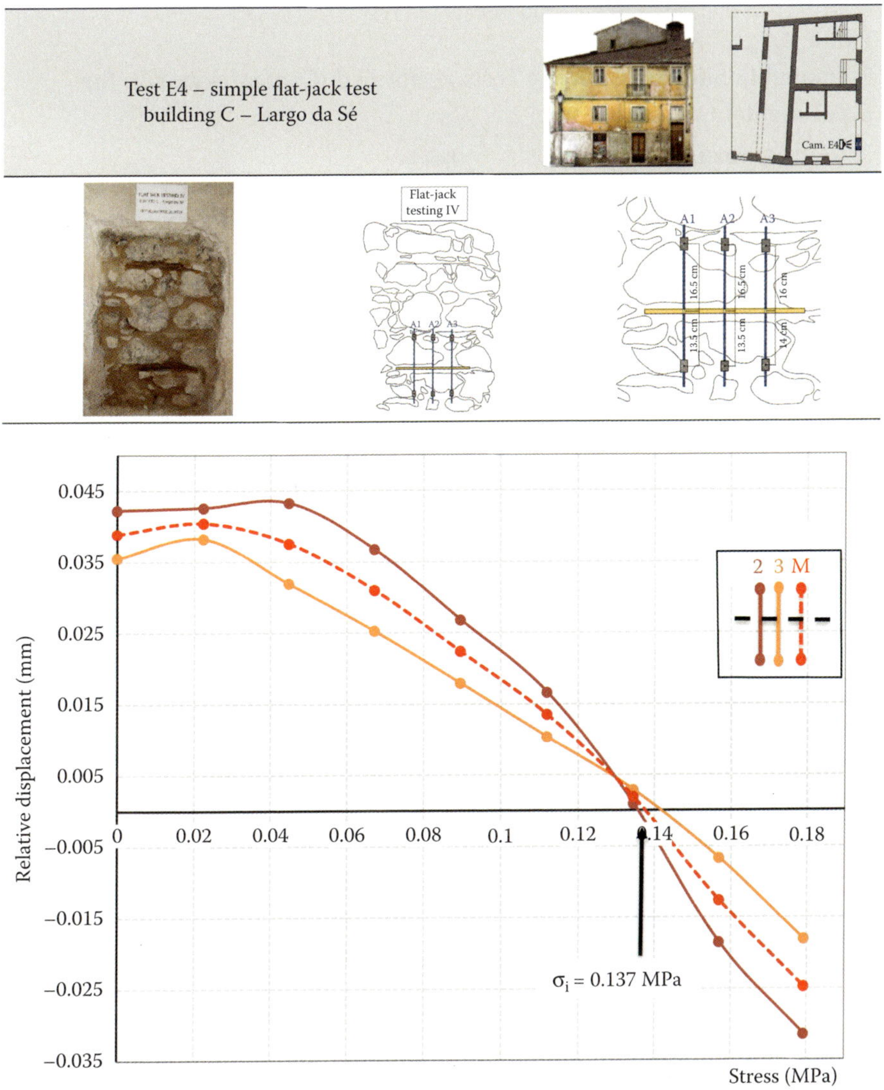

FIGURE 4.12 Single test E4: test representation and evolutions of displacements.

the scope of the study it was possible to characterise the different typology of the masonry walls that were found in the historical centre zone of Leiria. The walls are representative of the studies undertaken of the different buildings in the area. The research work developed suggests mechanical properties for masonry walls that can be used to support the parameters used in the realisation of rehabilitation projects of historical buildings, in this case the buildings of the historic centre of Leiria; namely, the modulus of elasticity and the maximum compressive stress.

TABLE 4.1
Estimated In Situ Stress for the Tests Performed (Comparison with the Experimental Obtained Value)

Test	Estimated Stress (MPa)	Obtained Stress (In Situ Test) (MPa)	Relative Variation (%)*	Observations
E1	0.107	0.087	+23	Very similar value, within the acceptable range of variance between ±24%
E2	0.110	0.113	−3	Very similar value, within the acceptable range of variance between ±24%
E3	0.194	0.158	+23	Very similar value, within the acceptable range of variance between ±24%
E4	0.175	0.137	+28	Very similar value, within the acceptable range of variance; the value is a little bit higher due to the % of mortar in the wall

* The acceptable relative variation range is ±24% according to ASTM.

FIGURE 4.13 Schematic protocol of double flat-jack tests: (a) opening the second cut; (b) cleaning and removing the dust inside the cut; (c) placement of the flat-jack; (d) setting and adjustment of potentiometers.

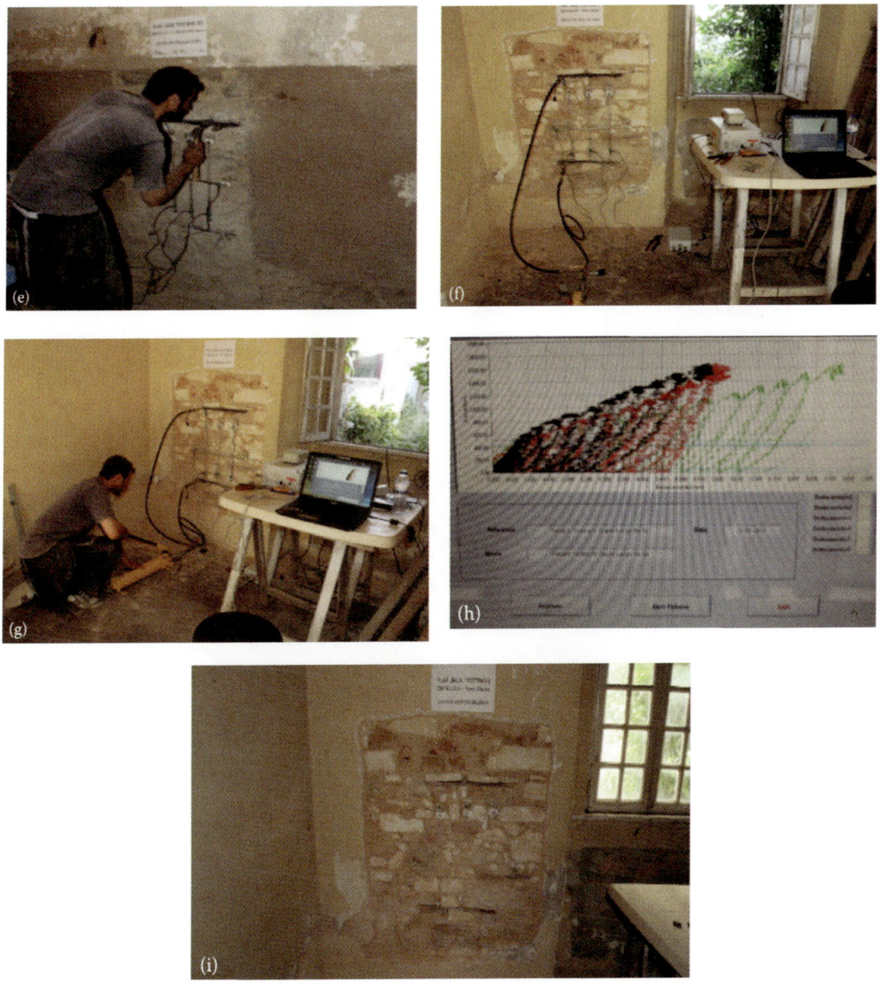

FIGURE 4.13 (Continued) (e) adjustments of the flat-jack; (f) connection of the whole system; (g) application of loading/unloading cycles; (h) global overview of the results; (i) the final appearance of the inspection window after the test.

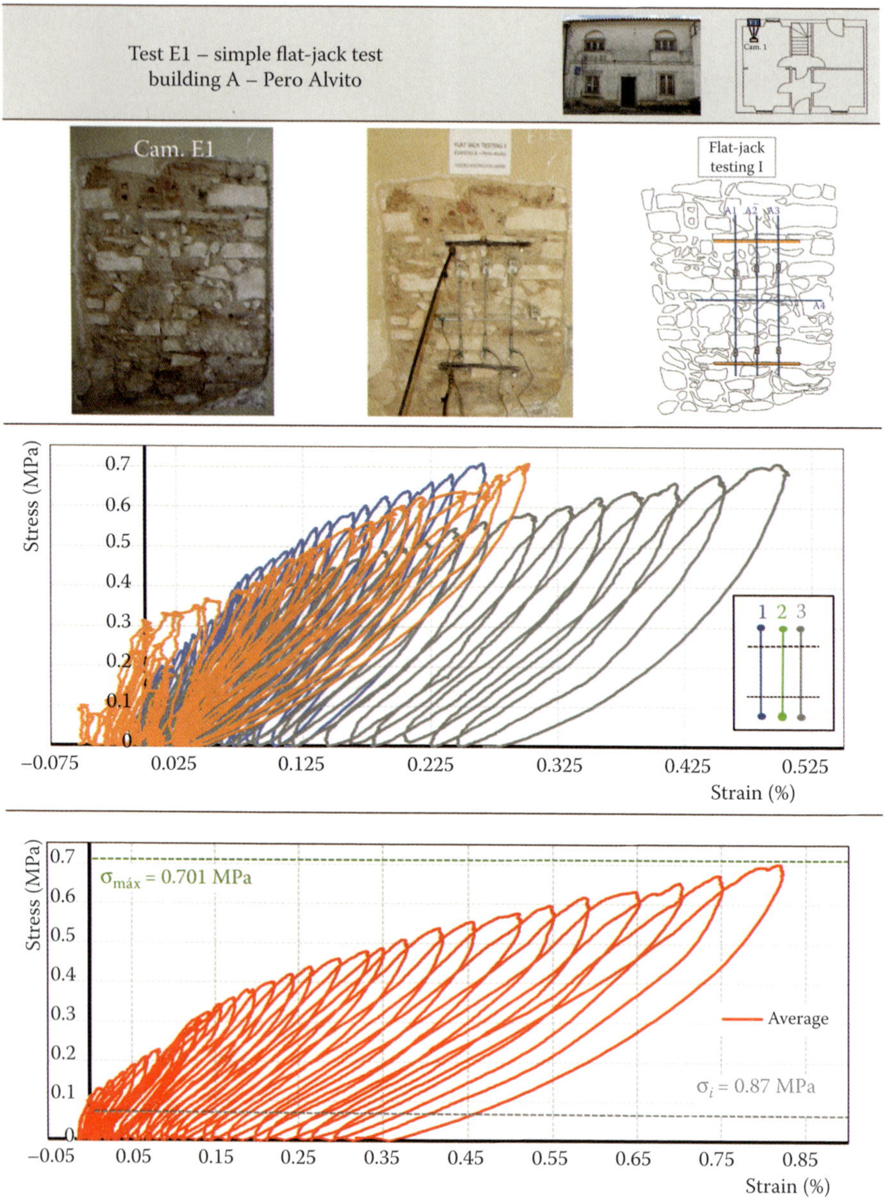

FIGURE 4.14 Double test E1: test representation and average stress-strain curves obtained from double tests.

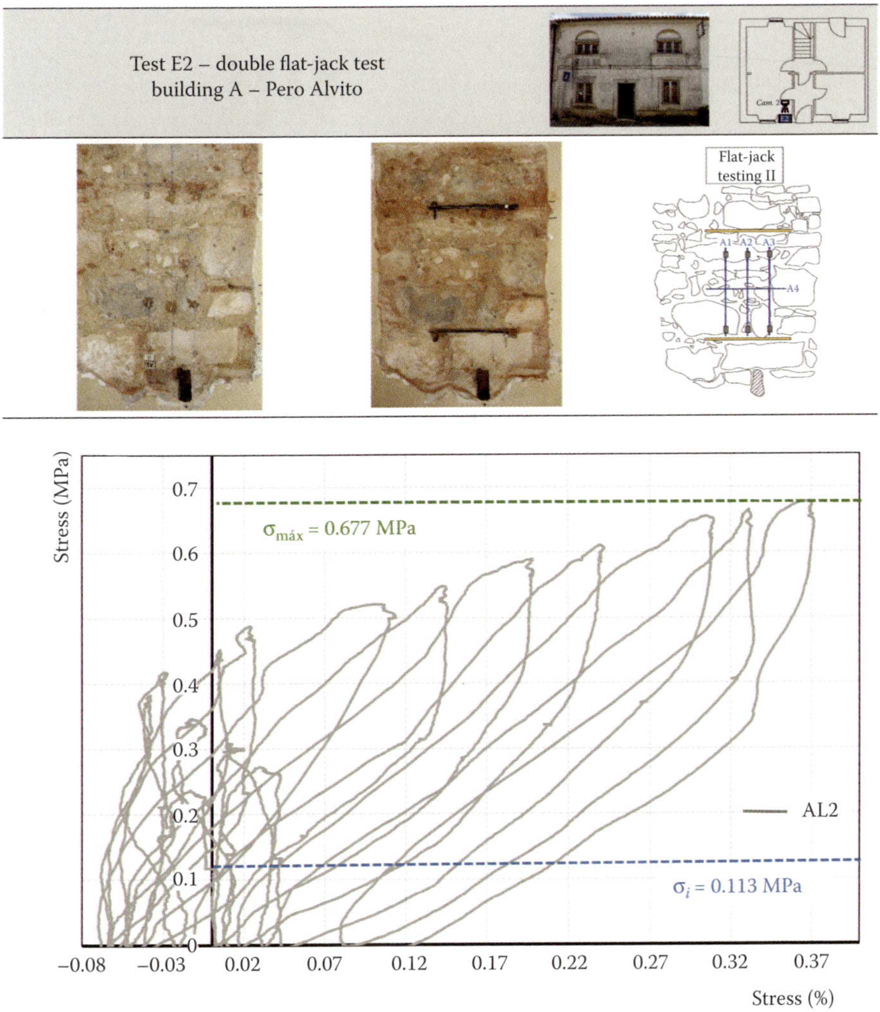

FIGURE 4.15 Double test E2: Test representation and average stress-strain curves obtained from double tests.

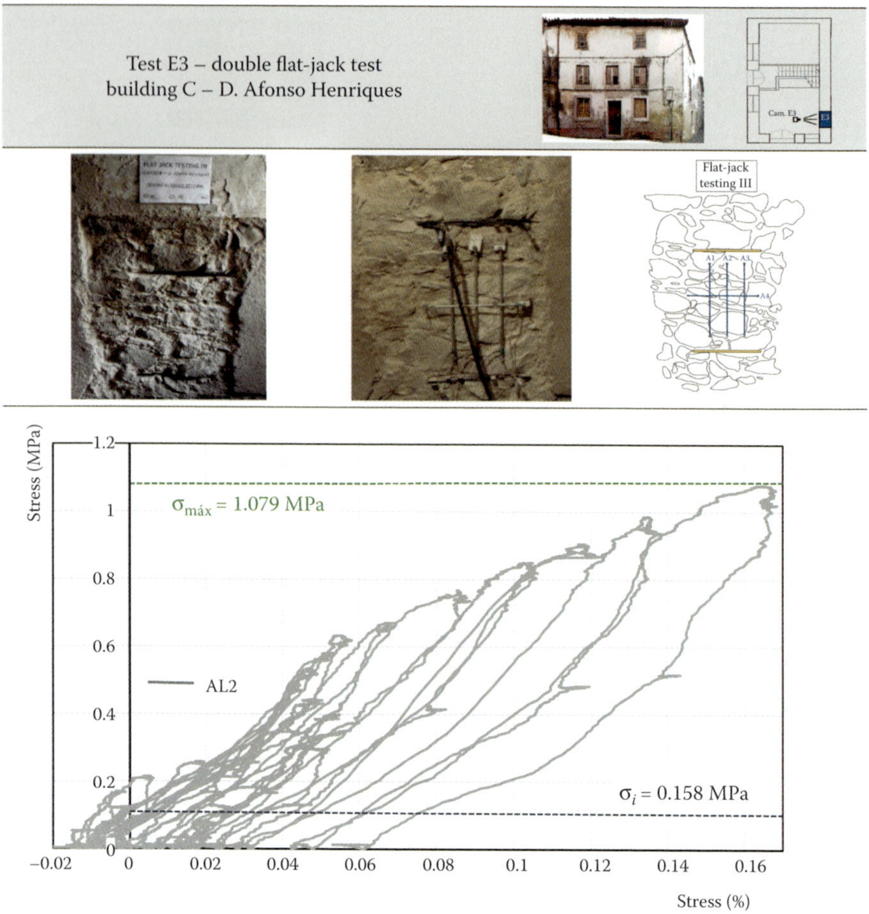

FIGURE 4.16 Double test E3: Test representation and average stress-strain curves obtained from double tests.

The installed stresses obtained in the simple flat-jack tests for the vertical loads show that, when compared with the maximum strength of the walls obtained with the double flat-jack, high safety factors (higher than 5) were observed and taken into account.

It was observed that variables such as typology, heterogeneity, age and quality of materials used, as well as the constructive method adopted, that is, the arrangement of the stones, are directly related to the final global behaviour of the masonry elements under study.

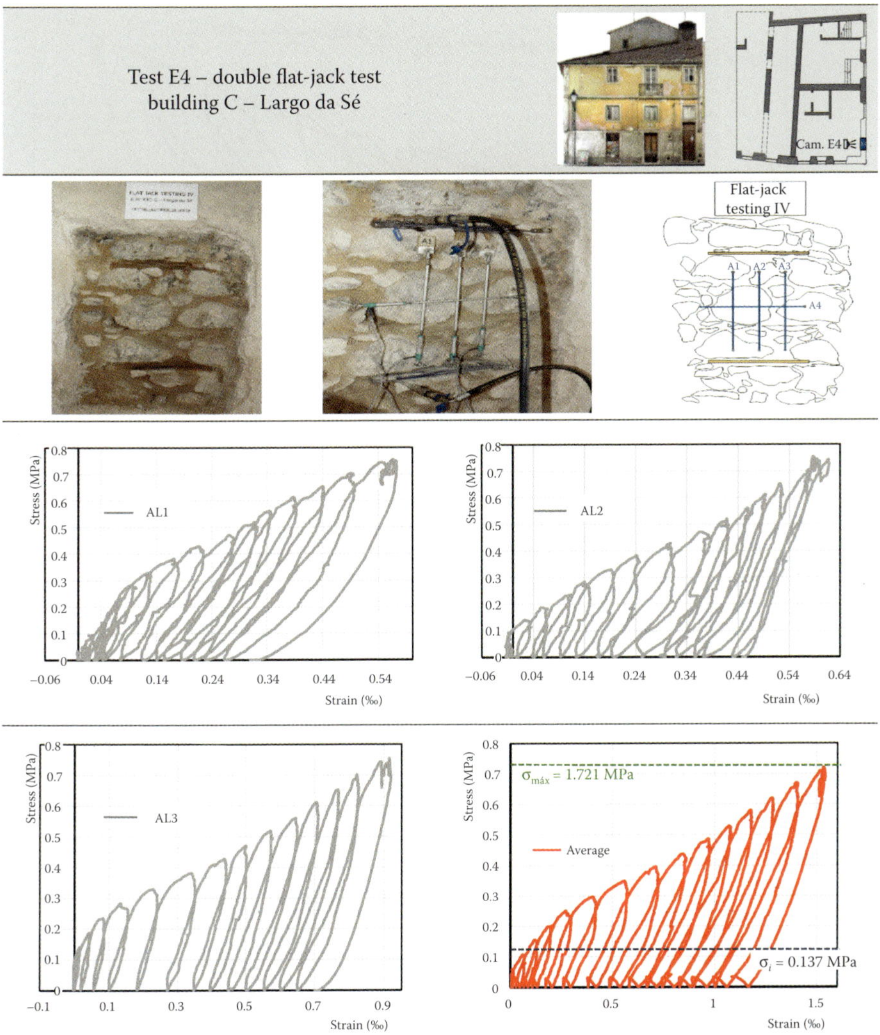

FIGURE 4.17 Double test E4: Test representation and average stress-strain curves obtained from double tests.

TABLE 4.2
Stress Results Obtained in Single and Double Flat-Jack Tests and Safety Ratio

Test	$\sigma_{in\ situ}$ (MPa) In Situ Stress	σ_{max} (MPa) Maximum Stress	$\sigma_{max}/\sigma_{in\ situ}$ Safety Ratio
E1	0.087	0.701	8.06
E2	0.113	0.677	5.99
E3	0.158	1.079	6.829
E4	0.137	0.721	5.263

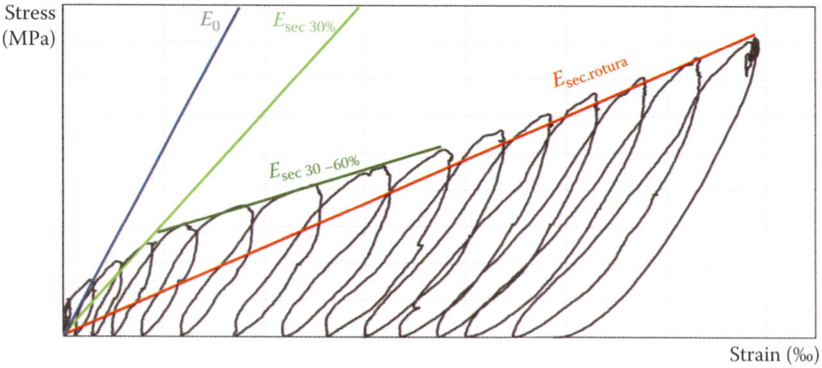

FIGURE 4.18 Different options for evaluation of the elastic modulus.

TABLE 4.3
Values Obtained for Different Elastic Modulus in Accordance with Figure 4.18

Test	E_0 (GPa)	$E_{sec30\%}$ (GPa)	$E_{sec30-60\%}$ (GPa)
E1	3.600	3.33	1.225
E2	6.198	5.35	0.590
E3	13.33	9.86	4.134
E4	3.010	1.01	0.350

REFERENCES

Almeida, C., J. P. Guedes, A. Arêde, C. Q. Costa, and A. Costa. 2012. Physical characterization and compression tests of one leaf stone masonry walls. *Construction and Building Materials* 30: 188–197.

ASTM. 1991. Standard test method for in situ measurement of masonry deformability properties using the flatjack method. In ASTM C1197-91. Philadelphia.

Binda, L., A. Saisi, and L. Zanzi. 2003. Sonic tomography and flat-jack tests as complementary investigation procedures for the stone pillars of the temple of S. Nicolò l'Arena (Italy). *NDT & E International* 36 (4): 215–227.

Binda, L. and C. Tiraboschi. 1999. Flat-jack test: A slightly destructive technique for the diagnosis of brick and stone masonry structures. *Restoration of Buildings and Monuments* 5 (5): 449–472.

Bosiljkov, V., M. Uranjek, R. Žarnić, and V. Bokan-Bosiljkov. 2010. An integrated diagnostic approach for the assessment of historic masonry structures. *Journal of Cultural Heritage* 11 (3): 239–249.

Carpinteri, A., S. Invernizzi, and G. Lacidogna. 2009. Historical brick-masonry subjected to double flat-jack test: Acoustic emissions and scale effects on cracking density. *Construction and Building Materials* 23 (8): 2813–2820.

Costa, C., D. Ribeiro, P. Jorge, R. Silva, A. Arêde, and R. Calçada. 2016. Calibration of the numerical model of a stone masonry railway bridge based on experimentally identified modal parameters. *Engineering Structures* 123: 354–371.

Gregorczyk, P. and P. B. Lourenço. 2000. A review on flat-jack testing. *Engenharia Civil UM* 9: 39–50.

Lourenço, P. B., F. Pagaimo, and E. Júlio. 2005. Caracterização das paredes de alvenaria da vila de Tentúgal (in Portuguese). *Revista Portuguesa de Engenharia de Estruturas* 54: 35–42.

Orbán, Z. and M. Gutermann. 2009. Assessment of masonry arch railway bridges using non-destructive in-situ testing methods. *Engineering Structures* 31 (10): 2287–2298.

Ranalli, D., M. Scozzafava, and M. Tallini. 2004. Ground penetrating radar investigations for the restoration of historic buildings: The case study of the Collemaggio Basilica (L'Aquila, Italy). *Journal of Cultural Heritage* 5 (1): 91–99.

RILEM. 2004. *In-situ Stress Tests based on the Flat-Jack*. Bagneux, France: RILEM.

Rocha, M., J. J. B. Lopes, and J. N. Da Silva. 1966. A new technique for applying the method of the flat jack in the determination of stresses inside rock masses. In *Proceedings of 1st ISRM Congress, International Society for Rock Mechanics*, 25 September–1 October 1966, Lisbon, Portugal.

Rossi, P. 1982. Analysis of mechanical characteristics of brick masonry tested by means of non-destructive in situ test. In *Proceedings of 6th International Brick and Masonry Conference*, Rome.

Simões, A., R. Bento, A. Gago and M. Lopes. 2016. Mechanical characterization of masonry walls with flat-jack tests. *Experimental Techniques* 40: 1163.

Simões, A., A. Gago, M. Lopes, and R. Bento. 2012. Flat-jack tests on old masonry buildings. In *Proceedings of 15th International Conference on Experimental Mechanics*, Porto/Portugal.

Vicente, R. S. 2008. Estratégias e metodologias para intervenções de reabilitação urbana: Avaliação da vulnerabilidade e do risco sísmico do edificadoda Baixa de Coimbra. PhD diss., Civil Engineering Department, University of Aveiro.

Vicente, R., T. M. Ferreira, J. A. R. Mendes da Silva, and H. Varum. 2015. In situ flat-jack testing of traditional masonry walls: Case study of the old city center of Coimbra, Portugal. *International Journal of Architectural Heritage* 9 (7): 794–810.

5 3D Structural Analysis to Support the Design Façade Retention System

Pedro Cardoso, João Veludo and Vitor Lopes

CONTENTS

5.1 INTRODUCTION

The rehabilitation of existing buildings in urban centres, with the demolition and/or renovation of their interiors, often requires the retention of their façades, especially in constructions of heritage value or in densely populated urban areas. This can be a complex and expensive process in many cases. Nowadays, there are several systems using different materials, such as steel or wood, and modular or particular systems designed for a specific case.

The purpose of this chapter is to present a 3D structural analysis to support the design of a façade retention system for an old three-storey building located in the historical centre of Leiria, Portugal. The building under study has a structure composed of masonry walls, wooden floors and roof, and wooden interior walls called *tabiques*.

The existing building is characterised by different conservation conditions from the point of view of the mechanical characteristics of the materials and the influence of the arrangement of the structural elements, such as walls, floors, roof and contiguous buildings, on the structural behaviour of the façade retention structure. The results of this study are compared with the dynamic analysis performed in situ with an accelerograph, thus enabling the calibration of the numerical model.

5.2 FAÇADE RETENTION SYSTEM FOR OLD MASONRY BUILDING

Most façade retention structures are erected as a prelude to a new scheme. However, there are cases where the façade requires urgent stabilisation following an earthquake, a fire, an explosion or similar extreme events.

The role of the planned retention structure is to provide the necessary support and restraint to the existing façade in the period between the beginning of the demolition work and the completion of the new interior structure. It is important to emphasise this duration to ensure that adequate restraint to the façade is provided at all times throughout the renovation period. To ensure stability and safety, the installation of the retention structure must start while the existing construction is still in place, before any existing restraint is removed. It must also remain, in whole or in part, until the new structure provides adequate restraint to the retained façade. During its lifetime, the retention structure must resist the forces applied to it and transmit these to the ground safely and without damage to the retained façade. The retention structure must be adequately (1) strong, (2) stable and (3) stiff. The retention structure should be idealised as a structure predominantly subject to lateral loading (Bussell et al. 2003, Gorse and Highfield 2009, Mahony 1996, Emmitt and Gorse 2014).

In emergency scenarios, particularly after an event of a seismic nature, the main objective of the façade retention systems is to sustain the façades, which may be weakened and/or may be liable to collapse onto adjacent streets. The main function of the façade retention structure is to control, or mitigate, the out-of-plane displacement of masonry walls while renovation operations are carried out, allowing the circulation of people and the intervention of local emergency agency authorities (FEMA-2A n.d., Grimaz et al. 2010a, Espasandin and Casas 2009, U.S. Army 2009, 2012). In this context, an assessment of the conservation status of the façade should be carried out, closely monitoring its seismic vulnerability to characterise the mechanisms of collapse (FEMA-1C n.d., Vicente et al. 2011, Bussell et al. 2003, U.S. Army 2009, 2012).

The evaluation of the mechanisms of collapse through numerical modelling provides an important contribution to the study of façade retention systems, enabling the forces transferred to the structural elements of those systems to be analysed. This evaluation is achieved through the consideration of the numerical modelling of the most common structural weaknesses in old buildings, namely, cracking in the corners and deterioration of wooden beams from the floor and roofs.

This chapter presents the numerical modelling from ongoing research, giving special emphasis to façade retention systems, for prevention and emergency situations in a masonry building that represents a widespread type of construction.

5.2.1 Vulnerability of Masonry Façade Walls

The evolution of the urban network in the historical centres of cities, due to the growth of rows of townhouses over the years, has its consequences both in structural

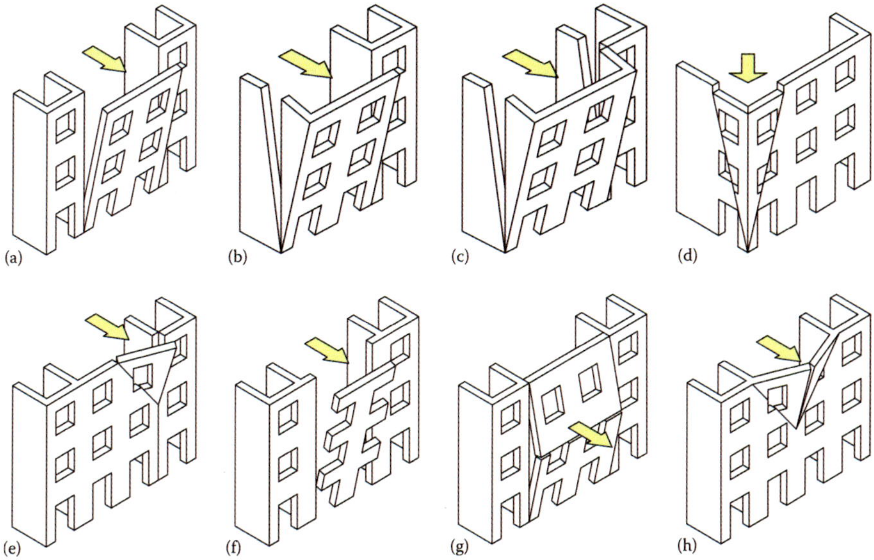

FIGURE 5.1 Collapse mechanisms. (Adapted from D'Ayala, D. and Speranza, E., *12th European Conference on Earthquake Engineering (12ECEE)*, 2002; Binda, L. and Saisi, A., *Structural Masonry*, 7(2), 85, 2005.)

terms and in terms of their seismic behaviour. Townhouse buildings often have problems in the connections between façades and orthogonal walls as well as between the walls of contiguous buildings. Furthermore, in the past, it was common to use the outer walls of contiguous buildings to support the floors and roofs of more recent buildings. The efficiency of the connections, especially between the façade walls and the orthogonal ones, is a vulnerable point, and they can collapse, partially or totally, under dynamic action.

Figure 5.1 illustrates some of the collapse mechanisms that have been identified in surveys of damage to buildings as a result of earthquakes that occurred in Italy and have been used in the assessment of the seismic vulnerability of masonry buildings based on limit analysis (D'Ayala and Speranza 2002, Binda and Saisi 2005).

From the analysis of the mechanisms of collapse, it is possible to observe that the location of the building in the urban network and the connections between structural elements (façade and inner walls and floors) have a significant influence on façades during the occurrence of an earthquake. The identification of these mechanisms for each case study is fundamental to the design of façade retention systems, as it allows the evaluation of the out-of-plane displacements as well as the determination of the forces transferred to the façade retention structure.

5.2.2 Types of Retention Systems of Existing Façades

There are several façade retention systems adopted for renovation and rehabilitation works or in emergency scenarios (e.g. the occurrence of an earthquake or a landslide). The choice of each façade retention system to be used is based on the state of conservation of the existing building and the surrounding areas. In emergency scenarios, façade retention systems are expected to fully or partially support façades that are damaged and/or could collapse onto the streets. The main function of the façade retention structures is to transfer horizontal actions caused by the weight of materials, imposed loads, wind or seismic action. Figure 5.2 shows several examples of façade retention systems based on the principles of load transfer to the ground and to the surrounding buildings (Grimaz et al. 2010a,b).

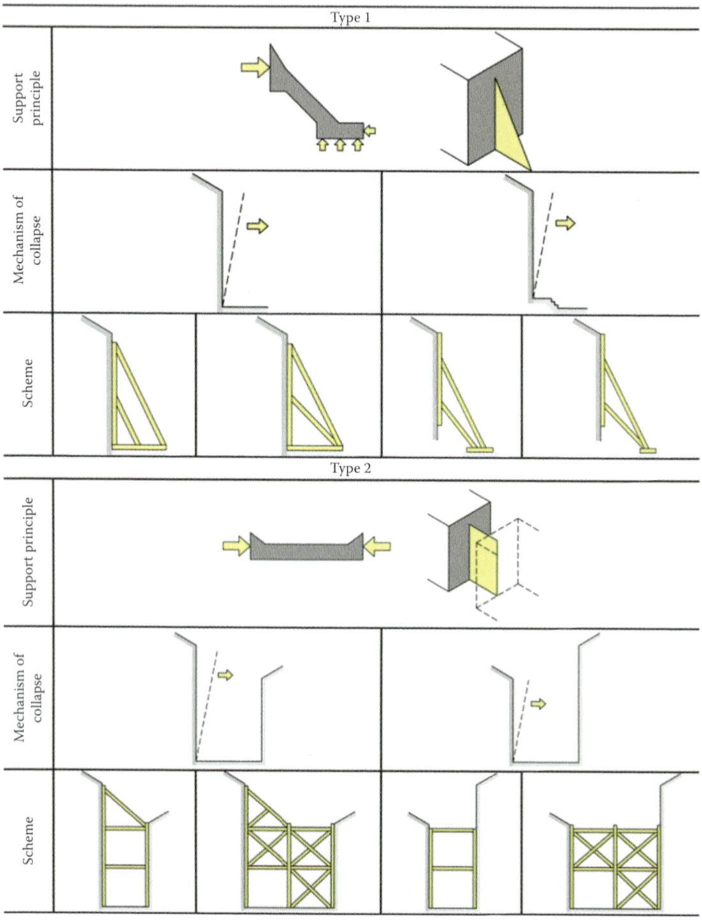

FIGURE 5.2 Types of façade retention systems. (Adapted from Grimaz, S. et al., *Manuale Opere Provvisional. L'Intervento Tecnico Urgente in Emergenza Sismica*, Ministero dell'Interno, Udine, 2010; Grimaz, S. et al., *Vademecun. Schede Tecniche delle Opere Provvisionali per la messa in sicurezza post-sisma da parte dei vigili del fueco*, Ministero dell'Interno, Udine, 2010.)

5.2.3 Structural Modelling

Finite element numerical models are currently one of the most commonly used tools in the analysis of old buildings. They enable continuous elements such as masonry to be modelled and can be carried out in two ways: (1) micro-modelling and/or (2) macro-modelling (Asteris et al. 2015, Costa et al. 2014, Lourenço et al. 1995, de Felice et al. 2017). In macro-modelling, simpler models can be used, assuming a linear elastic analysis, in contrast to micro-modelling which has to take into account structural discontinuities such as joints between masonry units. The latter is impractical and time-consuming in the study of old buildings with complex structures and geometries (Pulatsu et al. 2016).

For façade retention structures, the use of numerical models allows the study of the structural behaviour of the façades in the various phases of the constructive process of renovation or structural rehabilitation of the building.

In the next section, a case study is presented to illustrate the advantages of the use of numerical modelling in the design of façade retention structures.

5.3 CASE STUDY

5.3.1 Characteristics of the Building

The case study presented here is the Narciso Costa building, built in 1912, designed by the architect Ernesto Korrodi and located in the historical centre of Leiria (Figure 5.3). This building is composed of three floors and has a maximum height of façade equal to 11 m. The plan has a trapezoidal geometry with a 15×16 m^2 area. The main structure of the building comprises masonry walls, wooden floors, interior wooden walls (*tabique*) and a roof. The roof is made with trusses covered with tiles (type *marselha*). Structural walls are supported by stone masonry strip footings. In Table 5.1, a summary of the geometrical characteristics and material identification is presented.

5.3.2 Present Condition of the Building

During a visual inspection of the building, several structural and nonstructural pathologies were identified, the most relevant being (1) cracking in the corners of the building; (2) cracking between masonry walls and inner walls; (3) horizontal cracking in the connection between roof and masonry walls; and (4) moisture and deterioration in the connections of wooden beams and masonry walls.

5.3.3 Test for Measuring Environmental Vibration

Structural identification from environmental vibration testing was performed to better characterise the mechanical and structural properties, such as the modulus of elasticity of masonry walls in the three orthogonal directions (Figure 5.4a). The tests were performed with strong motion accelerographs. Readings were taken in various parts of the third floor where it was possible to place the accelerograph directly on the masonry wall (Figure 5.4b). In Table 5.2, the results obtained from the readings of environmental vibration are presented.

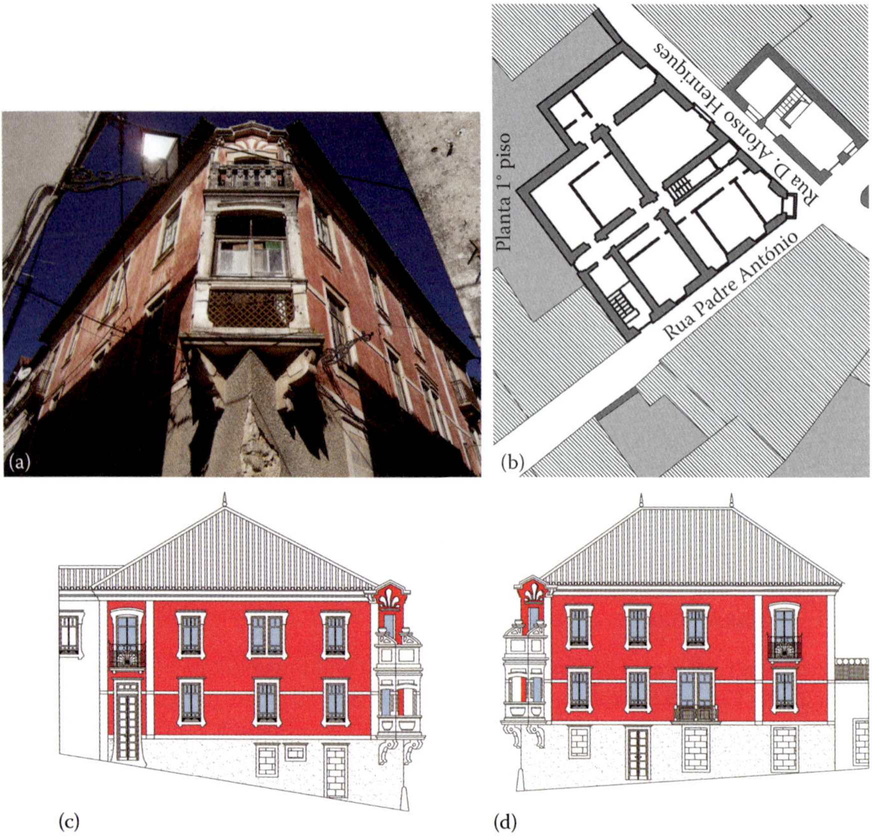

FIGURE 5.3 Building characteristics: (a) front view of the balcony; (b) plan view of the 1st floor; (c) view from Padre António Street; and (d) view from D. Afonso Henriques street.

TABLE 5.1
Characteristics of Structural Elements

Element	Material	Width (cm)	Height (cm)	Spacing (cm)
Stone masonry walls	Limestone	25–70	1100	–
Tabique walls	Wooden and mortar	11	330	–
Floors				
Wooden beams	Timber	10	16	45
Wooden boards	Timber	20	3	–
Roof (2 trusses)				
Wooden beams	Timber	10	16	45

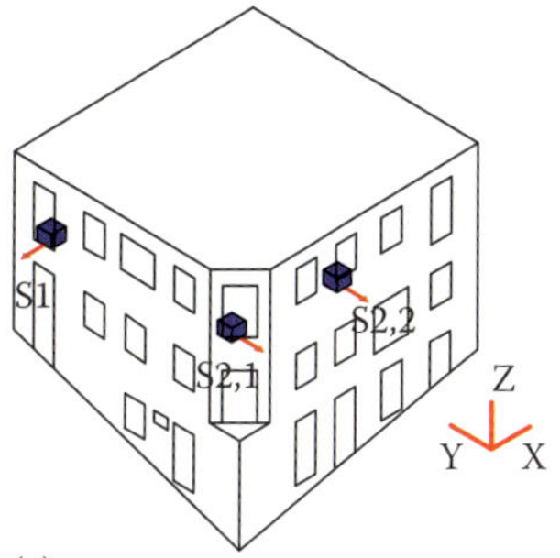

(a)

(b)

FIGURE 5.4 Environmental vibration test: (a) tests location; and (b) test S2.2.

TABLE 5.2
Summary of Environmental Vibration Tests

Position	Measured Frequency (Hz)	Average Frequency (Hz)
S1	8.12	8.66
S2.1	8.93	
S2.2	8.93	

5.3.4 Numerical Model

The numerical analysis was carried out using commercial software and a linear elastic analysis intended to obtain the forces acting on the adopted façade retention system, taking into consideration the mechanical properties of the materials, different boundary conditions and different states of conservation of the building. The geometry of the building has been defined through the analysis of the original architecture project and from the inspection campaigns carried out in situ. The structure was modelled using the following elements: (1) four node shell elements for masonry panels and inner *tabique* walls and (2) bar elements for wooden beams and wooden trusses of the roof. The walls were modelled with different thicknesses, taking into consideration walls with thicknesses of 25, 50, 65 and 70 cm. Adjacent to window openings, thick panels made of clay bricks (with a thickness of 19 cm) were used with the objective of assessing the influence that these elements might have on the overall behaviour. Floor and roof structures were modelled with articulated bar elements in the connection to the walls as well as with decorative elements. Pinned supports were considered for the foundations.

According to European norms as well as Portuguese National Annexes (EN-1990 2002, EN-1991-1-1 2002, EN-1991-1-4 2005, EN-1998-1 2004, EN-1998-1-3 2005) and Portuguese Standard (RSA 1983), both static and dynamic actions were considered: (1) permanent actions; (2) variable actions; and (3) seismic actions. The permanent actions considered were the self-weight and other permanent weights such as coatings in kitchens and bathrooms.

As variable actions, the wind and imposed loads for floors and roofs, according to the category of use, were considered. In accordance with Eurocode 8-part 1 (EN-1998-1 2004), the ground type considered was Type D.

According to the provisions of Eurocode 8-part 3 (EN-1998-1-3 2005), a behaviour coefficient equal to 1 was considered. This is a common situation in modelling masonry buildings with low ductility and low energy dissipation capacity (Vicente 2008).

For the mechanical characteristics of structural elements, the authors adopted values from the experimental laboratory, from field tests performed by other research reference studies, and from the literature (Vicente 2008, Farinha and Reis 1993, Rodrigues 2015): for masonry walls, a self-weight equal to 19.60 kN/m^3 and a Young elastic modulus equal to 1.75 GPa; for wooden elements (beams, boards, trusses and *tabique* walls), the corresponding values for class C18 (EN-338 2008).

According to the pathologies identified in Section 5.3.2, the authors have considered, in a numerical model, joint elements of negligible width and stiffness to simulate the cracks and the weaknesses detected in the corner connections between masonry walls. These elements also enable the mechanisms of collapse of the façades, specifically those associated with the out-of-plan deformations, to be simulated (see Section 5.2.2). These joint elements are shell elements of four nodes with a width of 5 cm each. The same elements were used in the connections between the masonry walls and the wooden beams of the floor.

Six different numerical models (Cases 1–6) were studied to assess the out-of-plane displacement for each of them and also to assess the load transfer in the two cases in which façade retention structures were added. In Figure 5.5a through 5.5f, the six different cases are presented.

All of the analysis was performed considering the XX direction and the combinations with the seismic action. For the façade retention structure, wooden elements of 10 and 24 cm in diameter, of class C18 (EN-338 2008), were adopted, considering the fact that these materials are easy to obtain and easy to work with and have adequate strength.

In Case 1 (Figure 5.5a), the original structure of the building was considered, including joint elements in the corners and also a decrease in the stiffness of these elements in the connections between wooden beams and masonry walls, to simulate the material deterioration and the weak tension resistance.

Concerning Case 2, only the façade walls were studied, without any horizontal restraint, simulating the constructive phases of a current method of rehabilitation that involves dismantling all the internal structural elements, leaving only the envelope walls, which are then replaced by reinforced concrete or steel structures, as illustrated in Figure 5.5b. This case also reproduces many real cases where the retaining structure is neglected due to cost management considerations.

Figure 5.5c shows Case 3, which considers the attachment of an isolated retaining structure as an autonomous structure with the function of transferring the loads to the ground. For the façade retention system, a 3D structure with vertical, horizontal and diagonal elements 10 and 24 cm in length was adopted. An internal portal frame had to be added to limit the lateral deflection at the top of the façade wall, as recommended by CIRIA 519 (Bussell et al. 2003), which limits this deflection to height/750. This consideration is aimed at increasing the stiffness of the retaining structure, avoiding displacements at the top of the wall, which could lead to collapse.

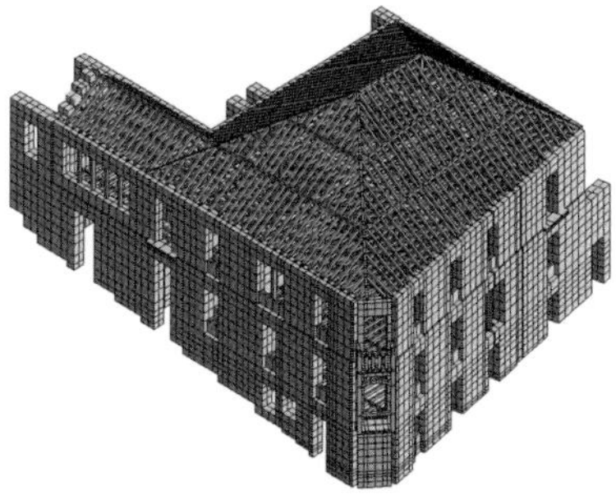

(a) Case 1 - Existing building in its actual state

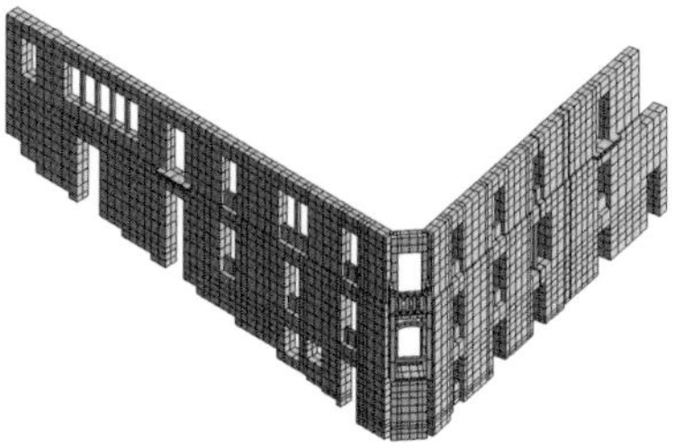

(b) Case 2 - Building without inner elements and without facade retention system

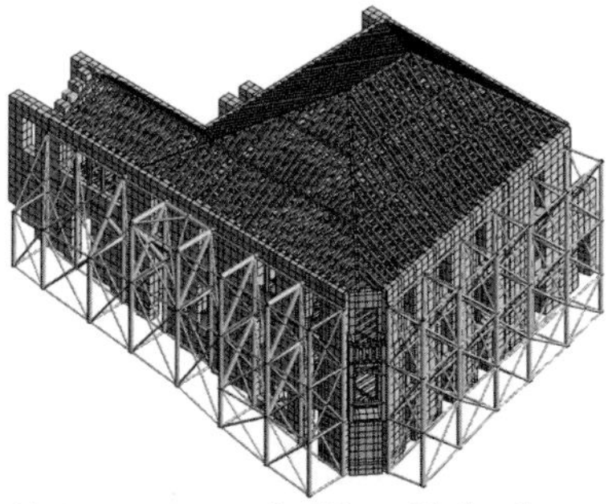

(c) Case 3 - Existing building with facade retention system

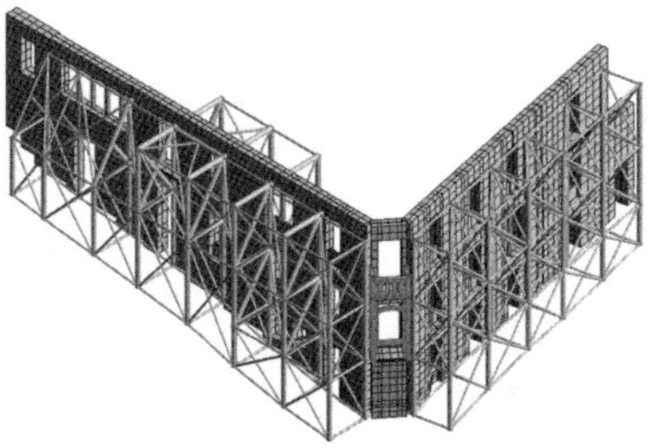

(d) Case 4 - Building without inner elements and with facade retention system

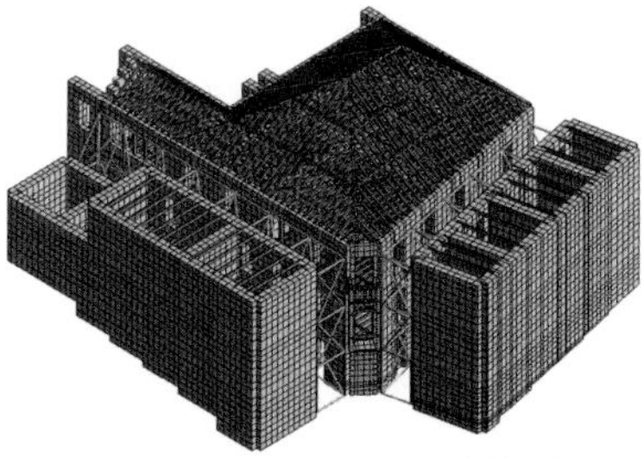

(e) Case 5 - Existing building with facade retention system supported in surrounding buildings

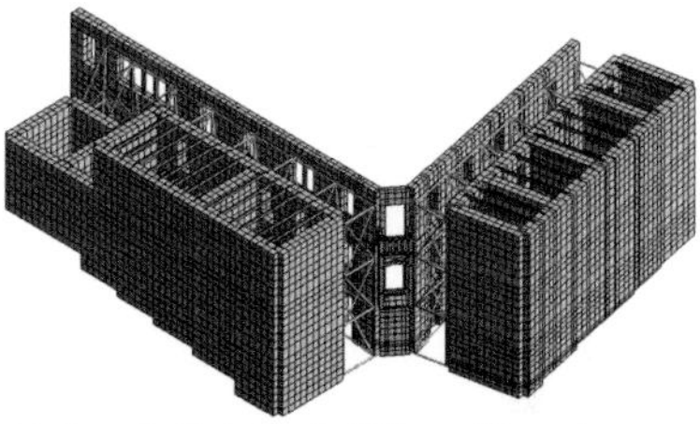

(f) Case 6 - Existing building without inner elements and with facade retention system supported in surrounding buildings

FIGURE 5.5 Finite element models.

The model presented in Case 4 is directly related to Case 2, but including the façade retention structure adopted in Case 3 (Figure 5.5d).

Based on Case 1, in Case 5, a faced retention structure with a structural scheme (illustrated in Figure 5.5e) was added to the model. To simulate the real supporting conditions of the retention structure, part of the surrounding buildings (across the street) were modelled. The retention structure is directly supported on the façades of the building at the floor levels. Such a structure requires an extensive survey on those buildings and their structural condition to avoid local or global damage. This type of structural solution also requires the permission of the owners of these buildings.

As presented in Case 5, the model shown in Case 6 compares the displacements of the façade wall when the support conditions of the retention structures are changed from an autonomous retention structure to one supported by surrounding buildings (Figure 5.5f).

5.4 ANALYSES AND DISCUSSION OF RESULTS

In Figure 5.6a, the first three modes of vibration, obtained from the numerical model used in Case 1, are presented. As can be observed, the frequency value obtained for the first mode (Figure 5.6a) is directly comparable to the values obtained from the environmental vibration tests (see Table 5.3), with an average relative error of 8.20%. The values obtained for the masonry walls of the building are within a range typically observed in structures of this nature (Vicente 2008).

In Figure 5.6b, the first shape modes associated with the out-of-plan deformation of the façades are presented.

The numerical model was calibrated based on the results obtained in environmental vibration testing as well as in the aforementioned pathologies.

The analysis of the previously mentioned maps reinforces the importance of the orthogonal walls in the structural behaviour of the masonry structures, thus demonstrating the need for shoring up these elements with façade retention structures, which control the out-of-plan deformation.

In Case 1, the importance of an efficient connection between the orthogonal walls and the façade to the local behaviour of the façade is clear; a large amplitude of

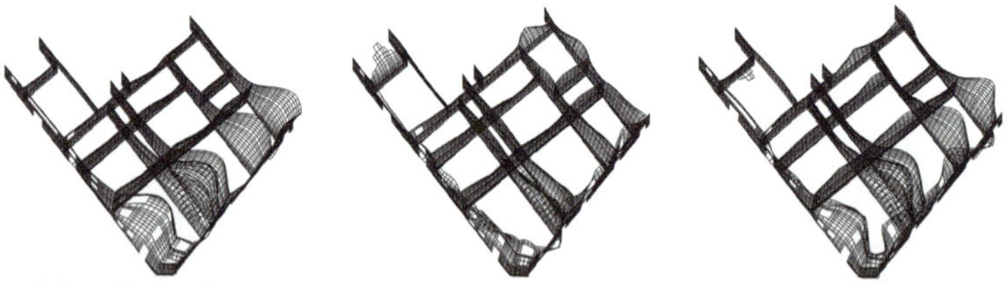

(a1) 1st Mode – f = 7.95 Hz (a2) 2nd Mode – f = 8.36 Hz (a3) 3rd Mode – f = 8.72 Hz

(a) Modal analysis

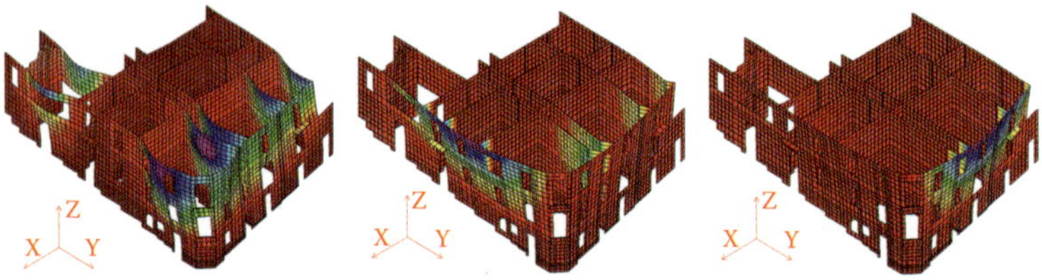

(b1) displacements in XX direction considering not fissured masonry max: 4.85mm; min: –0.01mm

(b2) displacements in XX direction considering fissured masonry max: 39.98mm; min: –0.03mm

(b3) displacements in YY direction considering fissured masonry max: 42.89mm; min: –0.03mm

(b) Case 1 –Displacement observed for seismic combination

FIGURE 5.6 Results from numerical model.

TABLE 5.3
Summary of Results Obtained for Different Cases

Case 1 and Case 2	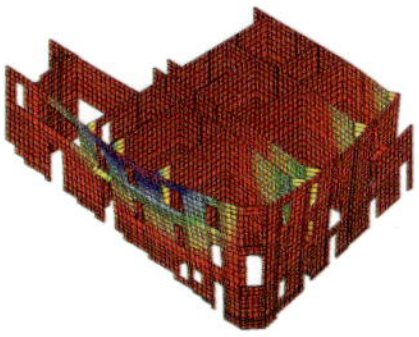	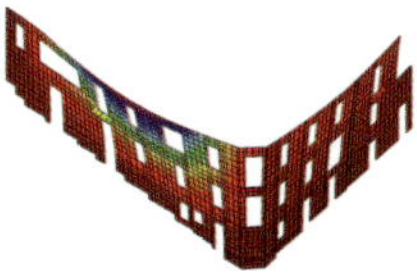
	Max: 39.98 mm; Min: −0.03 mm	Max: 46.76 mm; Min: 0.00 mm
Case 3	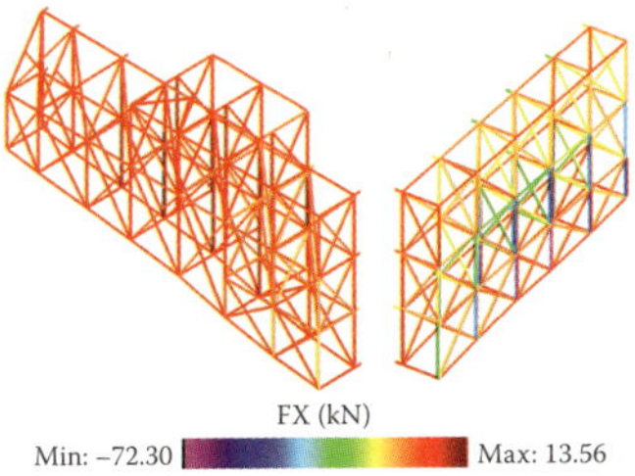	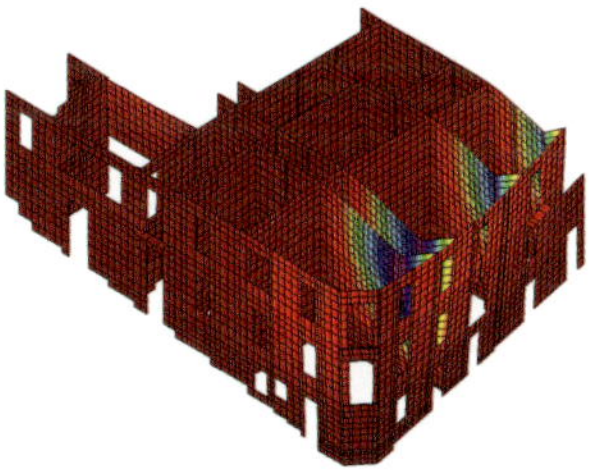
	Axial forces on the retaining structure	Max: 2.79 mm; Min: −0.10 mm
Case 4	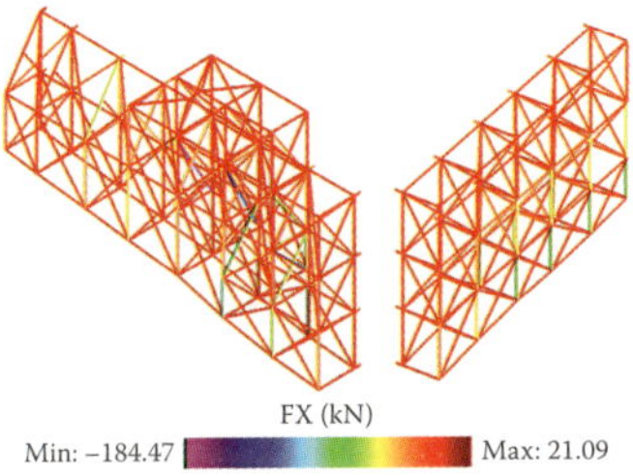	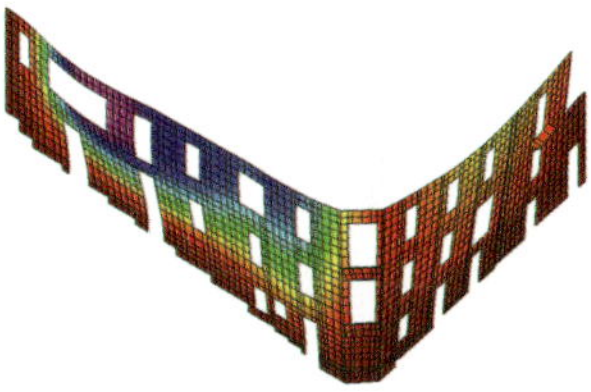
	Axial forces on the retaining structure	Max: 8.41 mm; Min: −0.02 mm
Case 5	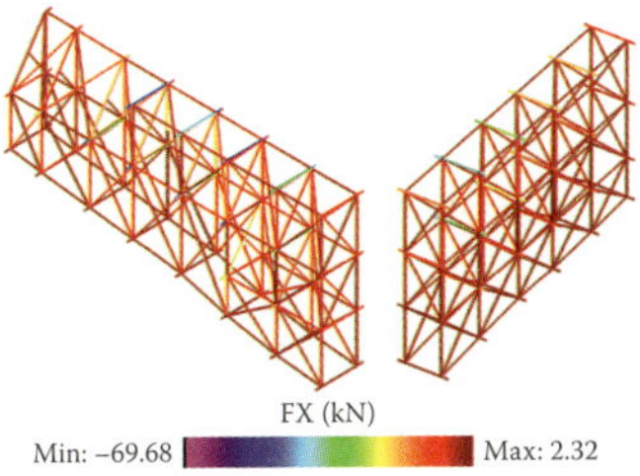	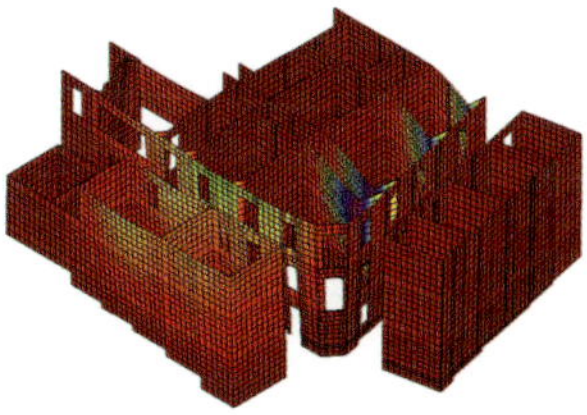
	Axial forces on the retaining structure	Max: 8.89 mm; Min: −0.04 mm
Case 6	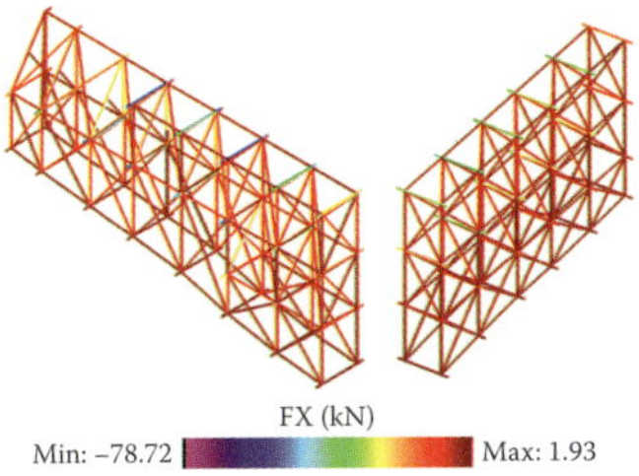	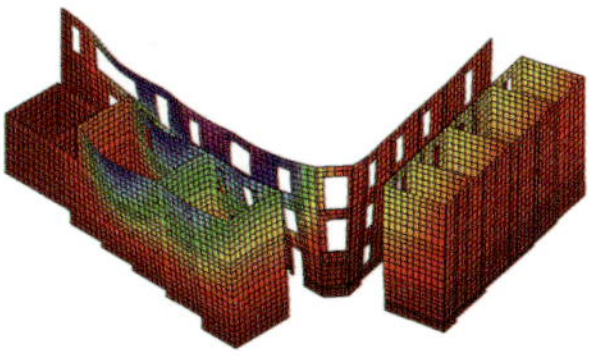
	Axial forces on the retaining the retaining structure	Max: 9.28 mm; Min: −0.03 mm

displacements is shown when these connections are not efficient. In the same way, displacements of internal walls parallel to the façade analysed have been observed. These displacements have been revealed in the form of effects on the façade through the diaphragm effect of the structure of the floor.

The results of Case 2 show the instability of the façade walls in an earthquake event when the inner structural elements are removed. Compared with Case 1, there is an increase in the amplitude of displacements, which may be explained by the resistance, however limited, of the elements of the joint.

The addition of the inner portal frame (Case 3) shows the importance that orthogonal walls can have for the behaviour of the structural scheme as a whole. In this case, wooden elements were anchored in the orthogonal walls, and displacements crossing these were transduced into small displacements at the top of the façade.

As expected, the results of Case 4 show a larger amplitude of displacements when compared with the previous case. This is supported by the fact that an autonomous retaining structure requires more inertia within the vertical elements of the portal frame than one that is attached to other elements, either surrounding buildings or inner walls, as in Case 3.

The results obtained in Case 5 show the influence of the retaining structure on the amplitude of displacements associated with the scenario suggested in Case 1, manifesting itself in a reduction of 22.4% in relation to the displacement of the initial case. It is also clear that the energy produced in the out-of-plane movement of the façade under study is not entirely dissipated within the elements that constitute the retention structure, thereby affecting its neighbouring buildings.

In Case 6, as observed in Case 5, the addition of the retaining structure controls the out-of-plane displacements, manifesting itself in forces transferred to the buildings that support it. In that case, the out-of-plane displacements have a reduction of 19.85%.

5.5 FINAL REMARKS

The previous sections represent some guidelines for the design of façade retention structures usually used in the rehabilitation and conservation of buildings and in prevention or emergency scenarios, showing the importance of numerical analysis, mainly modal analysis. They also show state-of-practice tools and proposals to solve current engineering problems related to the study of old masonry buildings.

The importance of a rigorous survey of the state of conservation of the structural elements of the building was emphasised, namely, the cracking of load-bearing walls and the degree of conservation of the wooden beams supporting the masonry. This allowed a more realistic calculation model that included the weaknesses of the façades and other structural elements, and hence, the design of a more optimised containment structure calculated for more realistic actions and conditions.

It was also verified that, for buildings with characteristics and sizes similar to the building studied, the state of conservation tends to define the conditioning actions for the design of the containment structures. In buildings in a reasonable state of conservation, such as the building studied, the seismic action is the conditioning. In buildings in an advanced state of degradation, the structural elements have a lower resistance capacity, and actions of lower intensity, such as gravitational and wind actions, can be determinant.

It was concluded that for the façade retention structure, wooden elements can be used, in the form of commercial sections and/or tree trunks, a natural and renewable material that is ecologically better than steel, which is most commonly used.

The current study also shows that, when allowed, the consideration of surrounding buildings is clearly a reliable option to support a façade retention structure. This façade support system may require an extensive structural survey of both the building supported and the building used as support, justified by the eventual damage that may occur in the supporting building, which may cause a chain reaction in the façade retention, leading to its collapse.

When an isolated retention structure is considered, stiffer structural elements may be needed. In extreme cases, this can lead to the use of mechanical elevation cranes and consequently, higher costs.

The corridor passage at the lowest level of the retaining structure is also a feature that has a direct influence on the façade displacements and stresses in the retaining structure. This problem also requires the adoption of larger wooden beam sections and stiffer connections between bars.

Further studies should be carried out to obtain optimal bar connections, with the purpose of achieving more secure and sustainable bars and connectors and an easier assemblage system, using preferably nonindustrialised solutions.

REFERENCES

Asteris, P. G., V. Sarhosis, A. Mohebkhah, V. Plevris, L. Papaloizou, P. Komodromos, and J. V. Lemos. 2015. Numerical modeling of historic masonry structures. In *Handbook of Research on Seismic Assessment and Rehabilitation of Historic Structures*, edited by Panagiotis G. Asteris and Vagelis Plevris, 213–256. IGI Global.

Binda, L. and A. Saisi. 2005. Research on historic structures in seismic areas in Italy. *Structural Masonry* 7 (2): 85.

Bussell, M., D. Lazarus, and P. Ross. 2003. *Retention of Masonry Façades. Best Practice Guide*. Vol. C579., CIRIA.

Costa, A. A., B. Quelhas, and J. P. Almeida. 2014. Numerical modelling approaches for existing masonry and RC structures. In *Structural Rehabilitation of Old Buildings*, edited by Aníbal Costa, João Miranda Guedes, and Humberto Varum, 285–305. Springer.

D'Ayala, D. and E. Speranza. 2002. An integrated procedure for the assessment of seismic vulnerability of historic buildings. In *12th European Conference on Earthquake Engineering (12ECEE)*, 9–13 September.

de Felice, G., S. De Santis, P. B. Lourenço, and N. Mendes. 2017. Methods and challenges for the seismic assessment of historic masonry structures. *International Journal of Architectural Heritage* 11 (1): 143–160.

Emmitt, S. and C. A. Gorse. 2014. *Barry's Advanced Construction of Buildings*. 3rd ed. Wiley Blackwell.

EN-338. 2008. Structural timber: Strength classes. European Committee for Standardisation (CEN).

EN-1990. 2002. Eurocode: Basis of structural design. European Committee for Standardisation (CEN).

EN-1991-1-1. 2002. Eurocode 1: Actions on structures. Part 1: EN 1991-1-1 Densities, self-weight and imposed loads for buildings. European Committee for Standardisation (CEN).

EN-1991-1-4. 2005. Eurocode 1: Actions on structures. Part 1: EN 1991-1-4 General actions. Wind actions. European Committee for Standardisation (CEN).

EN-1998-1. 2004. Eurocode 8: Design of structures for earthquake resistance, Part 1: General rules, seismic actions and rules for buildings. European Committee for Standardisation (CEN).

EN-1998-1-3. 2005. Eurocode 8: Design of structures for earthquake resistance, Part 3: Assessment and retrofitting of buildings. European Committee for Standardisation (CEN).

Espasandin, J. L. and J. I. G. Casas. 2009. *Apeos y Refuerzos Alternativos*. Munilla-Lería.

Farinha, J. S. B. and A. C. d. Reis. 1993. *Technical Tables*. P.O.B. (in Portuguese).

FEMA-1C. n.d. *Module 1C Structural Engineering Systems*. FEMA National US&R Response System Structural Collapse Technician.

FEMA-2A. n.d. *Module 2A Shoring Basics*. FEMA National US&R Response System Structural Collapse Technician.

Gorse, C. and D. Highfield. 2009. *Refurbishment and Upgrading of Buildings*. 2nd ed. Spon Press.

Grimaz, S., F. Barazza, M. Bellizzi, C. Bolognese, M. Cavriani, M. Caciolai, A. D'Odorico, et al. 2010a. *Manuale Opere Provvisional. L'Intervento Tecnico Urgente in Emergenza Sismica*. Ministero dell'Interno.

Grimaz, S., M. Cavriani, E. Mannino, L. Manuro, M. Bellizzi, C. Bolognese, M. Caciolai, A., et al. 2010b. *Vademecun. Schede Tecniche delle Opere Provvisionali per la Messa in Sicurezza Post-Sisma da Parte Dei Vigili del Fueco*. Ministero dell'Interno.

Lourenço, P. B., J. G. Rots, and J. Blaauwendraad. 1995. *Two Approaches for the Analysis of Masonry Structures: Micro and Macro-Modeling*. Delft University of Technology.

Mahony, B. 1996. Temporary works: The design of facade retention. Paper read at *Structures and Construction*. The Institution of Engineers of Ireland, 10 April, Dublin, Ireland.

Pulatsu, B., E. M. Bretas, and P. B. Lourenço. 2016. Discrete element modeling of masonry structures: Validation and application. *Earthquakes and Structures* 11 (4): 563–582.

Rodrigues, P. P. 2015. *Mechanical Characterization of Masonry Walls with Flat-Jack Tests in Leiria City Centre*. Polytechnic Institute of Leiria (in Portuguese).

RSA. 1983. *Safety and Actions Regulation for Buildings and Bridges Structures*. National Press (in Portuguese).

U.S. Army. 2009. *Field Operations Guide*. 2nd ed. US & R Essayons.

U.S. Army. 2012. *Field Operations Guide*. 7th ed. US & R Essayons.

Vicente, R. 2008. Strategies and methodologies for urban rehabilitation interventions. The vulnerability assessment and risk evaluation of the old city centre of Coimbra. PhD Thesis, University of Aveiro (in Portuguese).

Vicente, R., S. Parodi, S. Lagomarsino, H. Varum, and J. A. R. M. Silva. 2011. Seismic vulnerability and risk assessment: Case study of the historic city centre of Coimbra, Portugal. *Bulletin of Earthquake Engineering* 9 (4): 1067–1096.

6 3D Structural Analysis to Support the Rehabilitation of Old Masonry Buildings

Nelson Silva, João Veludo and Pedro Santos

CONTENTS

6.1 INTRODUCTION

The rehabilitation of old buildings is fundamental to ensure the preservation of existing cities and to promote their economic development, creating new opportunities for residential housing and commercial activities. This activity is critical when the existing construction was built using mainly materials such as masonry or timber instead of concrete and steel.

Old masonry buildings were built with structural stone exterior walls and wooden interiors with mortar walls, wooden floors and roofs covered with ceramic tiles. Steel members, usually disposed at floor levels, were implemented to increase the stiffness of the buildings. The foundations were frequently built using stone masonry footings with the same width as that of the base wall, or even wider. Wooden piles were frequently also adopted in areas with weak soils. In Portugal, a traditional wooden wall, commonly known as *tabique*, was developed for the interior walls.

The *tabique* walls, which were made up of vertical wooden boards and laths (horizontal wooden slats), were connected to the boards with steel nails. After this, the structure was covered by clay or mortar with very fine aggregates. *Tabiques* were intended to be used as structural and nonstructural members.

One of the main issues in the rehabilitation of old masonry buildings is the structural rehabilitation and the verification of the structural safety. The structural rehabilitation is driven by various circumstances, such as visible failure in structural members, change in use, or increased use, and even for seismic assessment and retrofit (Santos 2003).

In structural rehabilitation, certain aspects should be taken into account: namely, the architecture, the state of conservation, the existent structure and the historical values of the building.

Nowadays, the structural analysis of an old building is based on the modelling of its structural behaviour, and it is necessary to take into account the actual behaviour of the building when faced with various actions that are imposed on it, which should include static and seismic actions.

6.2 STRUCTURAL ANALYSIS AND MODELLING

The modelling of old structures presents a different challenge from the current (existing or new) structures of concrete or steel and becomes more complex due to the heterogeneous materials: in particular, the masonry and wooden walls that are composed of more than one material, the nonlinearity of the materials, and the anisotropy, which means different structural behaviour at different orientations of the loads (Roca et al. 2010, Asteris et al. 2015, Lourenço 2002).

Nowadays, there are several methods available for structural analysis, resulting in different structural approaches. Each approach is carefully selected and then adopted, all depending on the specific characteristics of each case study.

The adopted numerical models should always describe the original structure. When properly calibrated, these models will permit the identification of the theoretical damage produced by different types of action and the comparison of this with the actual damage observed in the structure (ICOMOS 2003). Therefore, a properly calibrated numerical model is a fundamental tool to ensure an adequate intervention. A numerical model of the existing structure, probably damaged, and of the strengthened structure will help the structural engineer to assess the current safety levels and to achieve the most economical intervention.

According to these statements, two different approaches can be used for the structural analysis of masonry structures: micro-modelling and macro-modelling. Macro-modelling can be used for the structural analysis of masonry structures built with thick walls, which will result in uniform stresses in those elements. This type of modelling is also faster and requires less memory and less complex finite element meshes, thereby presenting a lower cost. On the other hand, micro-modelling is used to understand the behaviour of the masonry structures, considering the heterogeneity and the nonlinear behaviour of the materials, and, when related to the seismic analysis, the various forms of mechanism failure (Lourenço et al. 1995).

A linear analysis can be used for macro-modelling when referring to the masonry work (stone blocks and mortars) as a homogeneous and continuous material, unlike micro-modelling, in which the properties of the materials are distinguished and should be taken into account separately (Lourenço 2002).

The structural behaviour mostly depends on the location of the building, the characteristics of the materials, the dimensions of the structure, the connections between the different members, and even the ground conditions (ICOMOS 2003).

For structural engineers, the assessment of the conditions of the existing structures (the identification of all the pathologies, including crack patterns and distribution), the geometrical survey, and the characterisation of the existent materials are crucial operations to support the data of the structural analysis (Binda et al. 2000). Whenever possible, the models should be calibrated and validated with experimental results that are obtained by tests conducted in situ and under laboratory conditions.

The current European design codes are aimed at the new concrete, steel, timber and masonry structures, with only additional measures for old buildings, which are composed of different materials. Furthermore, using the current codes may make it impossible to reach the safety levels that are required according to ultimate limit states.

The structural analysis and modelling of the case study presented in the next sections follow the guidelines and the principal concerns described in the previous sections.

6.3 CASE STUDY

6.3.1 Characteristics of the Building

The case study presented here is a building located in the old historical city of Leiria, in Tenente Valadim street, adjacent to the river Lis. The building, which was built in 1904, was designed by the architect Ernesto Korrodi. It is a residential building with three stories, with a trapezoidal shape, and the plan dimensions are 11.8×11.5 m^2. The main structural system of the building is comprised of masonry stone with mortar exterior walls, wooden floors, wooden interior walls and a roof. The roof is comprised of two wooden trusses (placed 3.35 m apart) covered by tiles (type *marselha*). The structural walls were sustained only by the use of stone masonry strip footings. Figure 6.1 illustrates the location, the façades, the plant, and the geometrical characteristics of the building and its structural members.

The analysis of the geometrical characteristics and the identification of the materials were performed by recording data for all the structural and nonstructural elements, and the results are summarised in Table 6.1.

6.3.2 Present Condition of the Building

In 2008, the building was subjected to nonstructural conservation and restoration work, which included the rebuilding of some damaged areas, the replacement of some damaged wooden elements and the treatment of all timber elements. All the structural members and the quality of the connections were visually inspected while

FIGURE 6.1 Building characteristics: (a) location; (b) View 1; (c) View 2; and (d) View 3.

the intervention was being done. It was possible to conclude that they were in a good condition, and only two existing corbel stones in Façade 3 (see Figure 6.1b) were replaced. A certain number of anomalies in the coatings and some noticeable deformations in the wooden floors were also identified. In the exterior and interior walls, a few days later, some cracking appeared due to the mortar shrinkage after the replacement of the coating.

6.3.3 Laboratory and Field Tests Performed on Interior and Exterior Walls

6.3.3.1 General

It is possible to assess the materials characteristically used in the construction of the building by conducting tests that are performed under laboratory conditions or directly on-site, adopting selected samples obtained from structural members, such

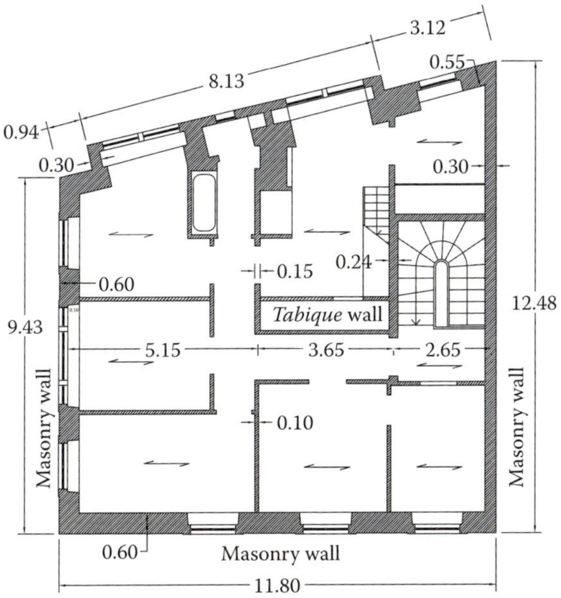

(e)

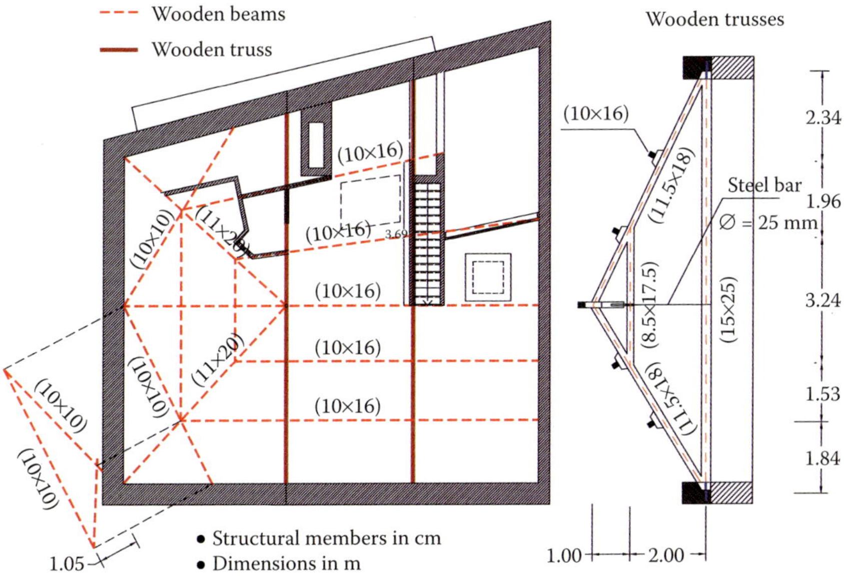

(f)

FIGURE 6.1 (Continued) (e) second floor plant layout; and (f) loft plant and truss layout.

TABLE 6.1
Characteristics of Structural Elements

Element	Material	Width (cm)	Height (cm)	Spacing (cm)
Masonry walls	Limestone	40–60		–
Tabique walls	Wooden and mortar	10–15	380	–
Floors				
Wooden beams		7.5	16	50
Wooden boards		20	3	–
Truss				
Bottom chord		15	25	–
Top chord		11.5	18	–
King post		11.5	18	–
Steel bar		2.5 (diameter)		
Collar tie		8.5	17.5	–
Purlins and ridge beam		10	16	

as cores for the masonry walls and small samples in the case of steel and timber elements.

For the masonry walls, the flat-jack test is one of the most common in situ tests, allowing the evaluation of the stress-strain relationship and of the modulus of elasticity. Mechanical laboratory tests can also be performed to determine the compression, tensile and shear strength of the interior wooden and even the stone walls.

6.3.3.2 Interior Walls (*Tabique* Walls)

In this work, a series of laboratory tests have been performed to evaluate the mechanical characteristics of the interior wooden walls. Three wooden *tabique* walls, with a 10 cm thickness and 1.9 m height, were fabricated using pine timber elements (strength class C18 according to EN 338) and mortar identical to the original ones. The height of the specimens was limited by the laboratory conditions to allow the fabrication, transportation and testing of the specimens.

To evaluate the axial and the bending stiffness, the specimens needed to be tested under compression. In Table 6.2, we have presented the geometrical characteristics of the wooden elements. It should be mentioned that the value of the thickness of the

TABLE 6.2
Geometrical Characteristics of *Tabique* Wall Elements

Element	Width (cm)	Thickness (cm)	Height (cm)	Spacing (cm)
Vertical boards	15	2.75[a]	190	6
Laths (horizontal slats)	2–3	2	100	3

[a] Average value

vertical boards shown in Figure 6.2 is a calculated value. In fact, the vertical boards used in the fabrication of the *tabique* walls are timber elements with live edges, with a difference in thickness between the top and the bottom of each element.

The filling and cover mortar was a commercial aerial lime mortar, provided by a local company, composed of an aerial lime with a sand aggregate mixed with water, which was one part lime for three parts of sand. This commercial mortar is a local product frequently used in the rehabilitation of old buildings due to its strong compatibility and aesthetic; it has a strong similarity to the existing mortar and masonry stone. Four different compositions were prepared, always changing the volume of water to achieve the appropriate consistency and workability. The volumes of mortar and aggregate were kept constant for each composition. To characterise the mortar lime, laboratory tests were performed: consistency, bulk density and air content. The tests performed followed the EN 1015 standard (EN 1015-3 1999, EN 1015-6 1998, EN 1015-7 1998). Table 6.3 summarises the results of the tests performed for each tested composition. The decision to select the mortar AL4 to be used in the fabrication of the *tabique* walls was based on the comparative analysis of the lime mortars, but also taking into consideration the pre-requirements initially defined.

In Figure 6.2 the different stages of the execution of the tabique walls are presented. First, the execution of the timber structure is shown. Each *tabique* wall is composed of six pine vertical boards with dimensions 27.5×150 mm^2 and a strength class C18 (EN 338) (Figure 6.2). Subsequently, the completion of the wall structure took place when the gaps that were more visible, at least between the laths and the vertical boards, were filled with lime mortar (Figure 6.2d), and then a metal wire mesh was placed and the cover applied (Figure 6.2e,f).

Figure 6.3a illustrates the adopted set-up test. Loads were applied using a hydraulic actuator of 1000 kN attached to a loading steel frame. For the displacement evaluation, linear variable differential transducers (LVDTs) were used. Two differential transducers were used for monitoring the axial deformation and another two for monitoring the out-of-plane deformation. A TML TDS 303 data logger recorded the data from the load applied and from the LVDTs used in the instrumentation. All the tests were conducted with load increments, with a load rate of 0.225 kN/s with a percentage of the estimated peak value (100 kN) (0%–20%Pu; 20%–40%Pu; 40% until failure).

Two different failure modes were observed in tests (Figure 6.3b): (1) failure of the vertical boards (observed in Wall 1) and (2) buckling failure (structural instability

TABLE 6.3
Summary of Test Results with Lime Mortar Specimens

Composition	Volume of Water (l/m^3)	Bulk Density (kN/m^3)	Air Content (%)	Consistency (%)
AL1	200	2053.83	4.2	20
AL2	225	2058.27	4.4	40
AL3	230	2074.13	4.7	50
AL4	235	2065.02	4.7	60

FIGURE 6.2 Preparation of *tabique* walls: (a) arrangement of timber elements; (b) lath geometry; (c) wooden structure; (d) mortar placement; (e) metal wire mesh placement; and (f) wall finished.

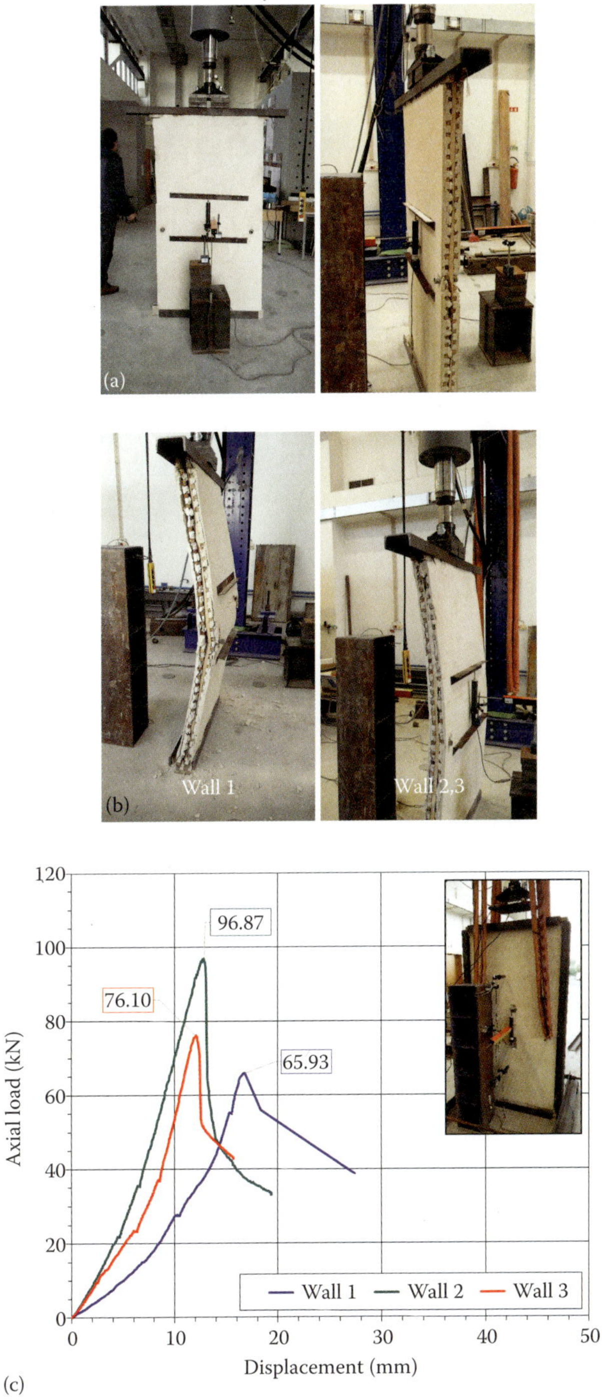

FIGURE 6.3 Set-up and results of performed tests: (a) test set-up; (b) modes of failure; and (c) load-displacement diagrams.

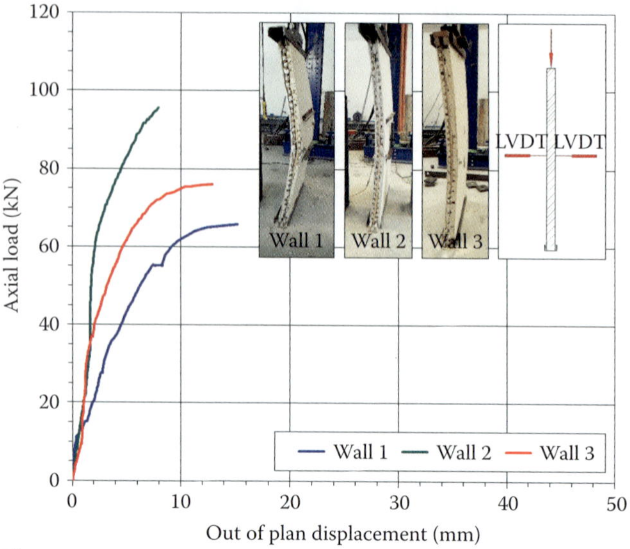

(d)

FIGURE 6.3 (Continued) (d) load-out of plan displacement diagrams.

TABLE 6.4
Summary of Tests Results with *Tabique* Walls

Specimen	Peak Load Pu (kN)	Displacement at Peak Load u_{Pu} (mm)	Axial Displacement u_a (mm)	Out-of-Plane Displacement u_t (mm)	Axial Stiffness *EA* (kN/m)	Bending Stiffness *EI* (kN.m²/m)
TW1	65.9	16.801	0.055	3.210	259,825	11.4
TW2	96.9	12.840	0.050	1.725	372,094	16.6
TW3	76.1	12.070	0.050	1.940	287,547	13.2
Average	79.6				306,489	13.7

observed in Walls 2 and 3). During the tests, it was noticeable that the lime mortar cover had behaved properly.

Figure 6.3 illustrates the axial load-displacement curves for all the compression tests. It can be seen that the obtained curves have a similar shape in all the tests. Each curve presents two different stages. First, a linear elastic branch is identified, presenting each curve in three slightly different values of tangent stiffness. After the peak load is achieved, a sudden drop in the wall capacity can be observed.

In Table 6.4, the results obtained in the compression tests are displayed. The axial (u_v) and the transverse displacement (u_t) values are registered at 50% of the peak load. The axial stiffness (*EA*) was calculated considering the average axial displacement that was registered between two reference points (Wall 1: 0.46 m; Wall 2: 0.41 m; Wall 3: 0.40 m). The bending stiffness (*EI*) was calculated after considering the value registered due to unplanned displacement in the initial stage of the wall, which was loaded within the limits of elastic deformation (50% of the peak load).

The values obtained for the axial and the bending stiffness were compared with the theoretical; these values were calculated by taking into consideration only the vertical board elements. The axial and the bending stiffness values were estimated as follows:

Axial stiffness

$$u_a = (P \cdot l)/(EA) \tag{6.1}$$

Bending stiffness

$$1/r = u_t \cdot \pi^2 / H^2 \tag{6.2}$$

$$1/r = M/(EI) = (P \cdot u_t)/(EI) \tag{6.3}$$

where:

u_a	is the axial displacement
P	is the applied load
l	is the distance between two reference points
u_t	is the out-of-plane displacement
E	is the modulus of elasticity
A	is the cross section
$1/r$	is the curvature
H	is the height of the wall
I	is the moment of inertia

An estimated value for the axial and the bending stiffness can be calculated considering the cross section and the moment of inertia of the six vertical board elements. The real value for the modulus of elasticity is parallel and equal to 9 GPa. The values obtained are 272,500 kN/m for the axial stiffness (considering the wooden elements and a mortar with a modulus of elasticity of 1 GPa) and 14.04 kN.m^2/m for the bending stiffness. These values are close to those achieved from previously performed tests.

The results obtained in this series of laboratory tests showed the variation that can occur in terms of the material properties and mechanical behaviour of the *tabique* walls. This can be seen in Figure 6.2; the vertical boards used in each *tabique* wall do not have a regular shape and are different from wall to wall. This situation leads to different values in the geometrical characteristics of the vertical timber elements and to different axial and bending stiffness in each wall.

6.3.3.3 Masonry Stone Walls

The stress-strain relationship of the masonry stone walls could not be characterised by the means of in situ tests due to restrictions imposed by the owner of the building. Nevertheless, a series of tests performed under laboratory and in situ conditions

TABLE 6.5
Summary of Test Results with *Masonry Stone* Walls

Tests	Ultimate Stress σ_u (MPa)	Modulus of Elasticity E (GPa)
Flat-jack (in situ)	0.70–1.00	0.35–1.20
Flat-jack (in laboratory)	0.77 (avg. value)	2.23 (avg. value)
Compression tests (in laboratory)	0.99 (avg. value)	0.36 (avg. value)
Ambient vibration tests		0.46–1.02

Source: Pinheiro, P. Structural characterization of masonry walls from the Historical Center of Leiria. MSc diss. Polytechnic Institute of Leiria (in Portuguese), 2015.

enabled the mechanical behaviour of the masonry walls of the old and historical city of Leiria to be characterised (Pinheiro 2015). The laboratory tests consisted of the execution of two stone masonry walls using the same techniques and materials as adopted in the construction of the existing buildings in the old city. These specimens were tested under compression to breaking point. Flat-jack tests were also performed under laboratory and in situ conditions to evaluate the existing stress-strain state in the wall and to obtain the axial and the bending stiffness of stone structural walls.

In Table 6.5, a summary of the characteristics of the walls and the most important findings is presented.

Table 6.3 shows the difference in values obtained for the modulus of elasticity from tests performed in situ and in the laboratory.

6.4 3D AND NUMERICAL MODELLING

Based on the assessment of the construction geometry, material characteristics and the geometry of the structural members, the authors developed a 3D geometrical model and a 3D structural model of the building. The results of the 3D modelling are shown in Figure 6.4a.

The structural stability of the building was analysed using Robot Structural Analysis software. To perform the structural analysis, a linear elastic material that measures behaviour was incorporated. Due to the objectives of the analysis and to the uncertainty related to parameters such as soil and the walls' mechanical characteristics, a nonlinear numerical analysis (Lourenço 2002) was not necessary.

Special attention was given to stress distributions and deformations, as well as stress concentrations in critical zones (joints, section variations, etc.). The static loads considered were dead weight, coatings and imposed loads, coatings, and the dynamic loads were the seismic action, evaluated according to the European safety code (EN 1998-3 2005).

The adjacent building was not seen as an influence on the model presented in Figure 6.1b, even though it presents a similar geometry and mass structure and was built using the same materials and founded on the same soil.

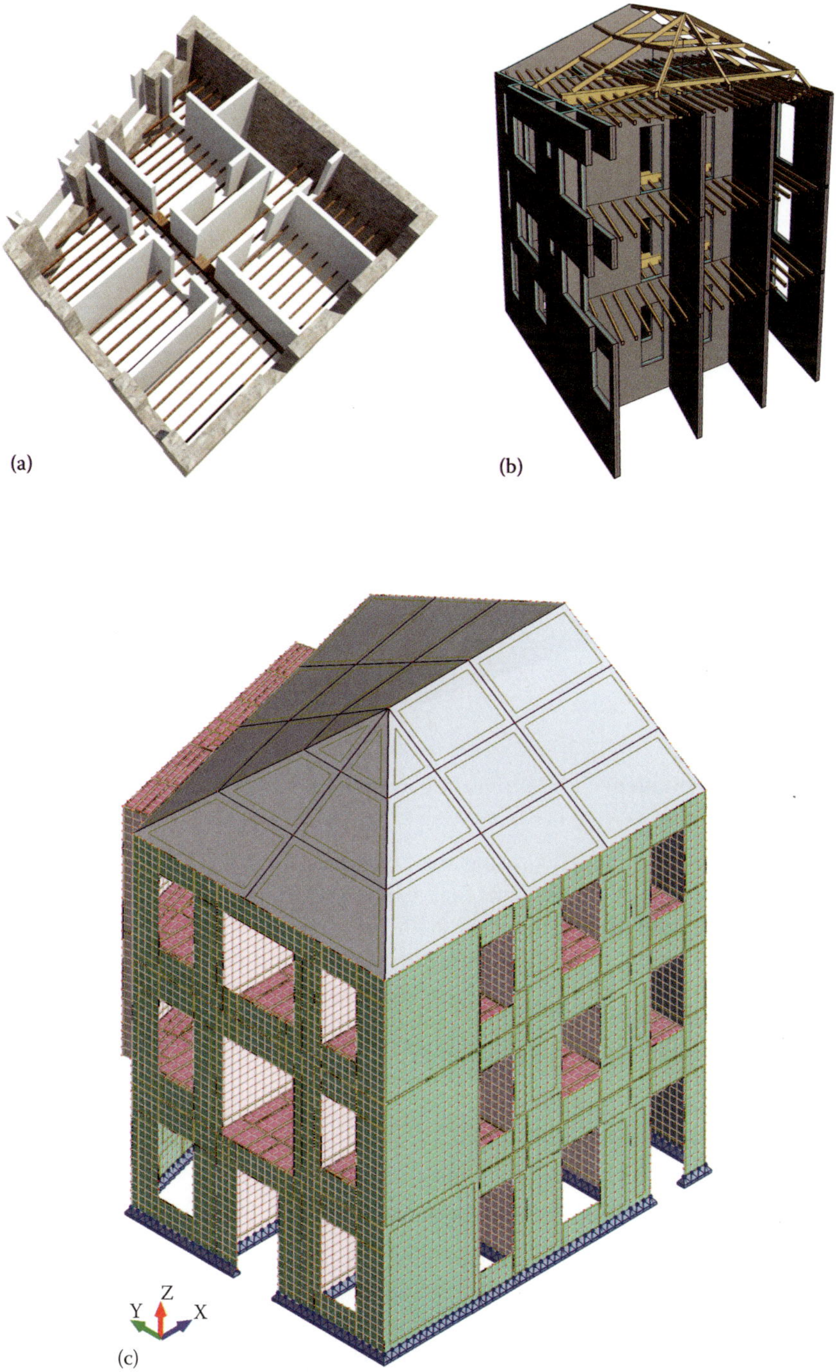

FIGURE 6.4 3D and numerical modelling of the building: (a) and (b) 3D geometrical modelling, (c) 3D numerical model.

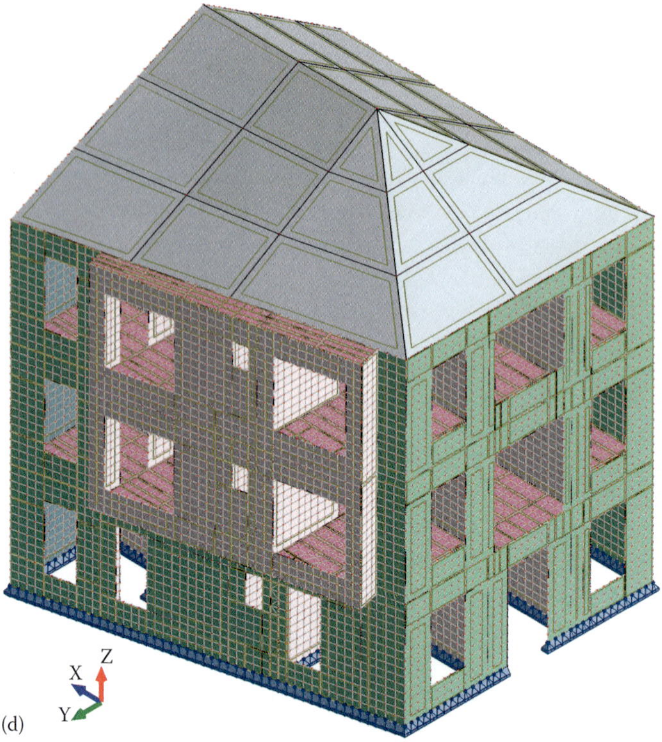

(d)

FIGURE 6.4 (Continued) (d) 3D numerical model.

The structure was modelled using shell elements for walls; beam and shell elements for floors; and beam elements for trusses (Figure 6.4).

According to the tests performed with the *tabique* walls, and based on the research conducted by Pinheiro (2015), the physical and mechanical properties of the materials used in the numerical model are presented in Table 6.6.

According to the EN 1998-3 (2005) standard, the behaviour factor for horizontal seismic actions was considered equal to 1.0.

TABLE 6.6
Physical and Mechanical Properties of the Materials Used in the Numerical Model

Element	Material	Modulus of Elasticity (GPa)	Poisson Ratio	Specific Weight (kN/m³)
Masonry walls	Limestone	1.5	0.26	22.6
Tabique walls	Wooden and mortar	9.0	–	15.5
Wooden elements		9.0	–	6.0

In the absence of more accurate evaluations, both contributions to stiffness may be accepted as one-half of their respective values (EN 1998-3 2005).

6.4.1 Structural Assessment

In Table 6.7, the values of frequency and period are presented for the first nine modes of vibration obtained from the modal analysis. The contribution of each mode (mass) to the global response of the building is also given. The modal analysis was performed for a total of 150 modes, but due to the similarity among the majority of them, it was decided to present only the most relevant results here.

Figure 6.5 gives the main results of the structural analysis. Figure 6.5a presents the most relevant of the three vibration modes: Figure 6.5a1, third mode shape corresponding to the YY direction; Figure 6.5a2, fifth mode shape corresponding to the XX direction; and Figure 6.5a3, seventh mode shape corresponding to the YY direction. These results showed that the building presents an adequate behaviour when subjected to seismic loads without a significant torsion effect. Moreover, the large number of vibration modes obtained, each corresponding to a lower mass, is due to the independent vibration of each *tabique* wall. Further research is necessary to investigate the behaviour of the joints between them and the wooden floors.

By analysing the stress distributions for each load case and the load combinations, it can be concluded that, in terms of principal pressure, the stress level is low to moderate and, in general, of very little importance. In the masonry walls, the highest compression stress is 0.22 MPa, and the tension stress is equal to 0.17 MPa, which does not indicate cracking (Figure 6.5). As seen in Figure 6.5, the maximum stress levels were identified near the joints and section variations, but this is all related to the type of analysis undertaken. Mesh refining or the adoption of advanced analysis methods could help to improve the stress distributions in these areas.

TABLE 6.7
Summary of Modal Analysis Results

Mode	Frequency (Hz)	Period (s)	Rel.mas.UX (%)	Rel.mas.UY (%)	Rel.mas.UZ (%)
1	2.64	0.38	0.02	0.24	0.00
2	2.65	0.38	0.25	0.26	0.00
3	*2.73*	0.37	6.81	*14.78*	0.00
4	2.75	0.36	6.86	14.90	0.00
5	*2.87*	0.35	*53.85*	18.17	0.02
6	2.99	0.33	53.85	18.18	0.02
7	*3.08*	0.32	60.49	*50.50*	0.03
8	3.26	0.31	60.59	50.64	0.03
9	3.31	0.30	60.59	50.64	0.03

Note: Rel. mass is the effective modal mass, and the italic values are the values corresponding to the modes presented in Figure 6.5.

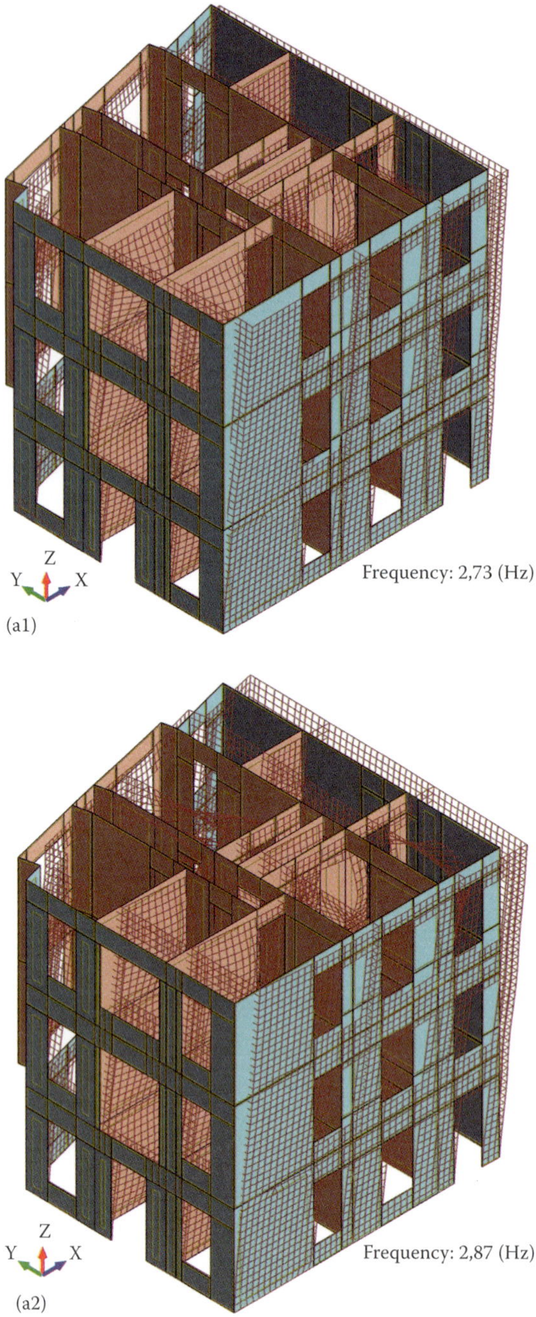

FIGURE 6.5 Results from structural modelling. (a) Modes of vibration. (b) Principal stresses in masonry walls. (c) Principal stresses in Tabique walls. (d) Global displacements for seismic action. (e) Deformation in wooden floors.

FIGURE 6.5 (Continued) (a) Modes of vibration.

The maximum global displacements and interstorey drift observed (Figure 6.5) are also reduced and compatible with the adjacent buildings and within the limits of the values proposed by the Eurocode 8-part 1 (EN 1998-1 2004). Visual inspections did not reveal any damage in the façades.

6.5 FINAL REMARKS AND INTERVENTIONS FOR STRUCTURAL IMPROVEMENT

The rehabilitation of old buildings is fundamental to ensure the preservation of existing cities. Only with this type of intervention is it possible to promote their sustainable development and to create new opportunities for residential housing and commercial activities.

Old buildings were mainly built using masonry and timber, instead of concrete and steel. Therefore, current design codes and standards are not fully adequate to perform an analysis on their structural behaviour. In many cases, the analysis of old buildings following the guidelines of the current codes could result in the identification of serious structural problems, which in some cases would not be true.

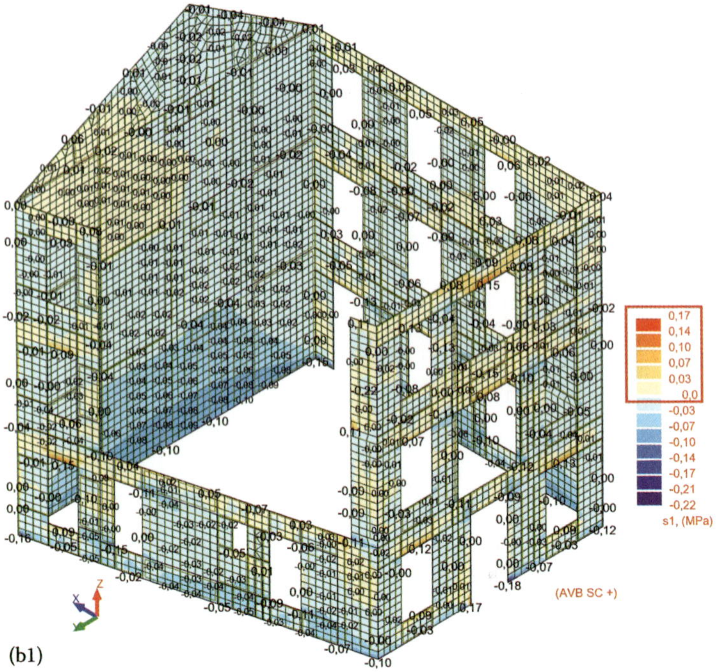

(b1)

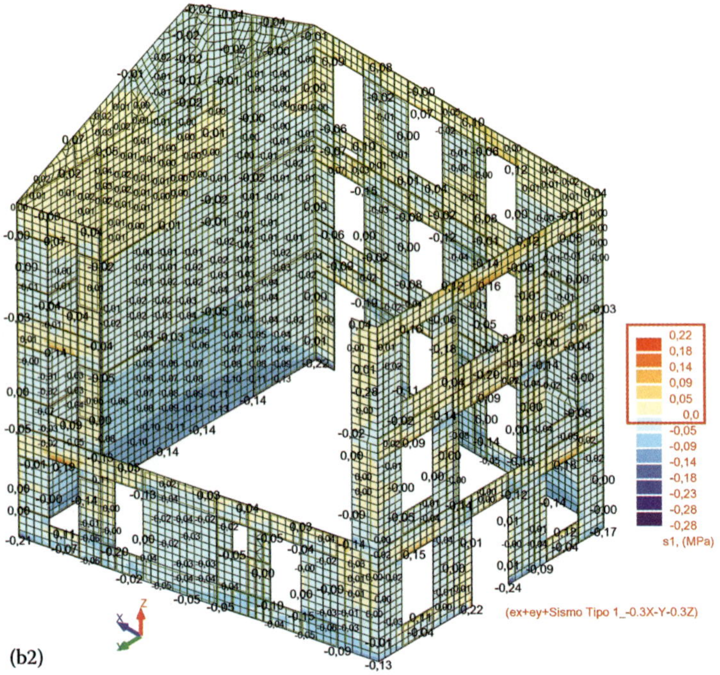

(b2)

FIGURE 6.5 (Continued)

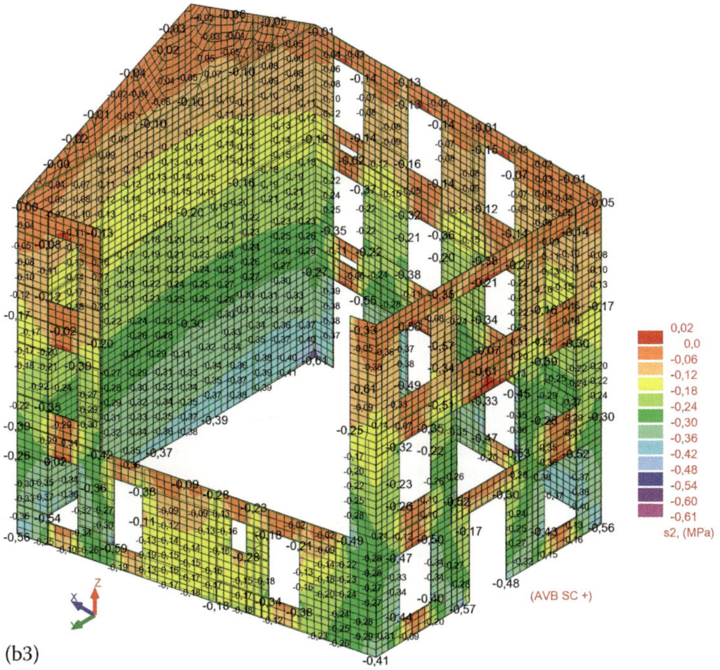

(b3)

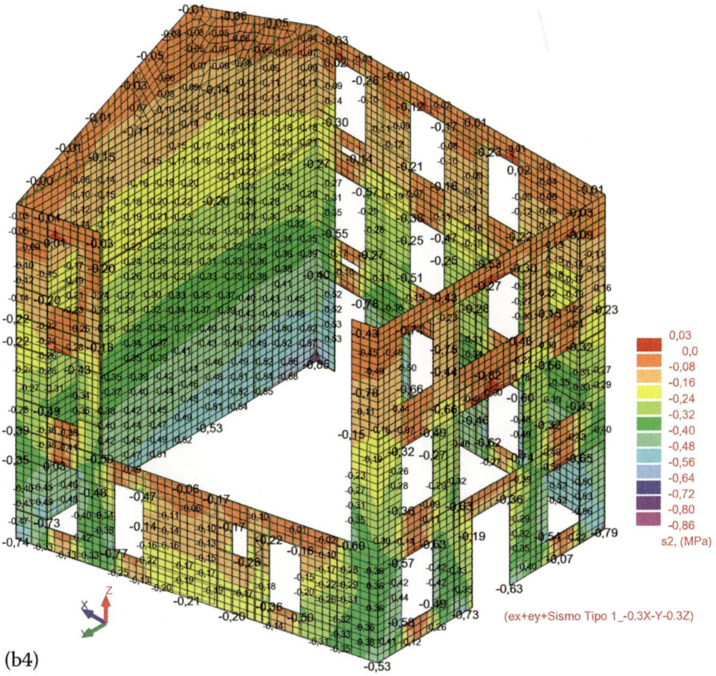

(b4)

FIGURE 6.5 (Continued) (b) Principal stresses in masonry walls.

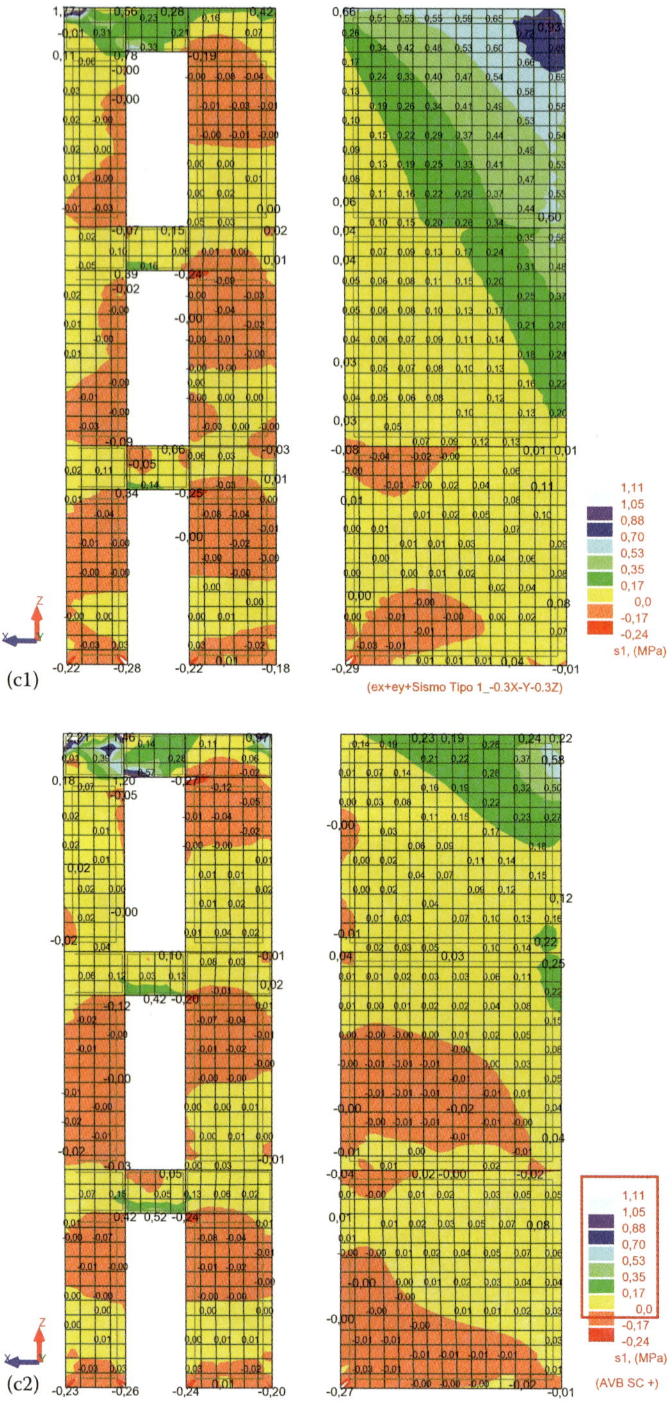

FIGURE 6.5 (Continued) (c) Principal stresses in Tabique walls.

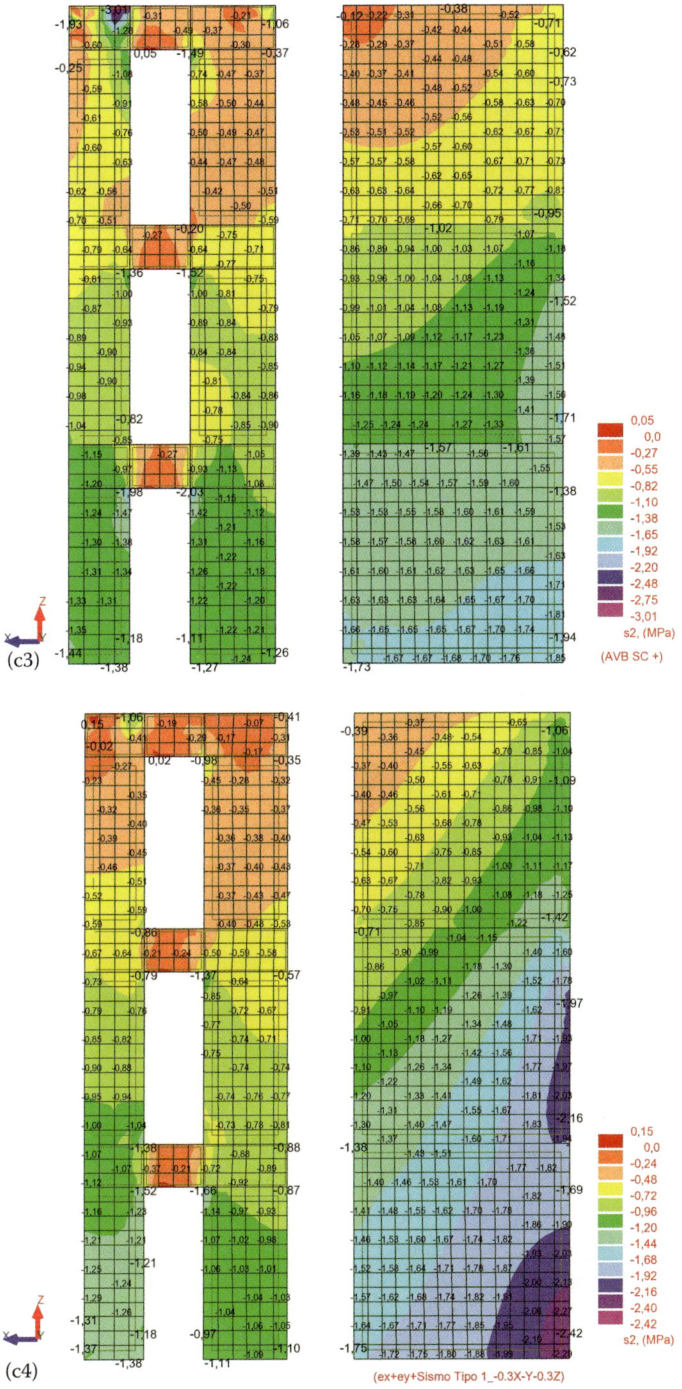

FIGURE 6.5 (Continued)

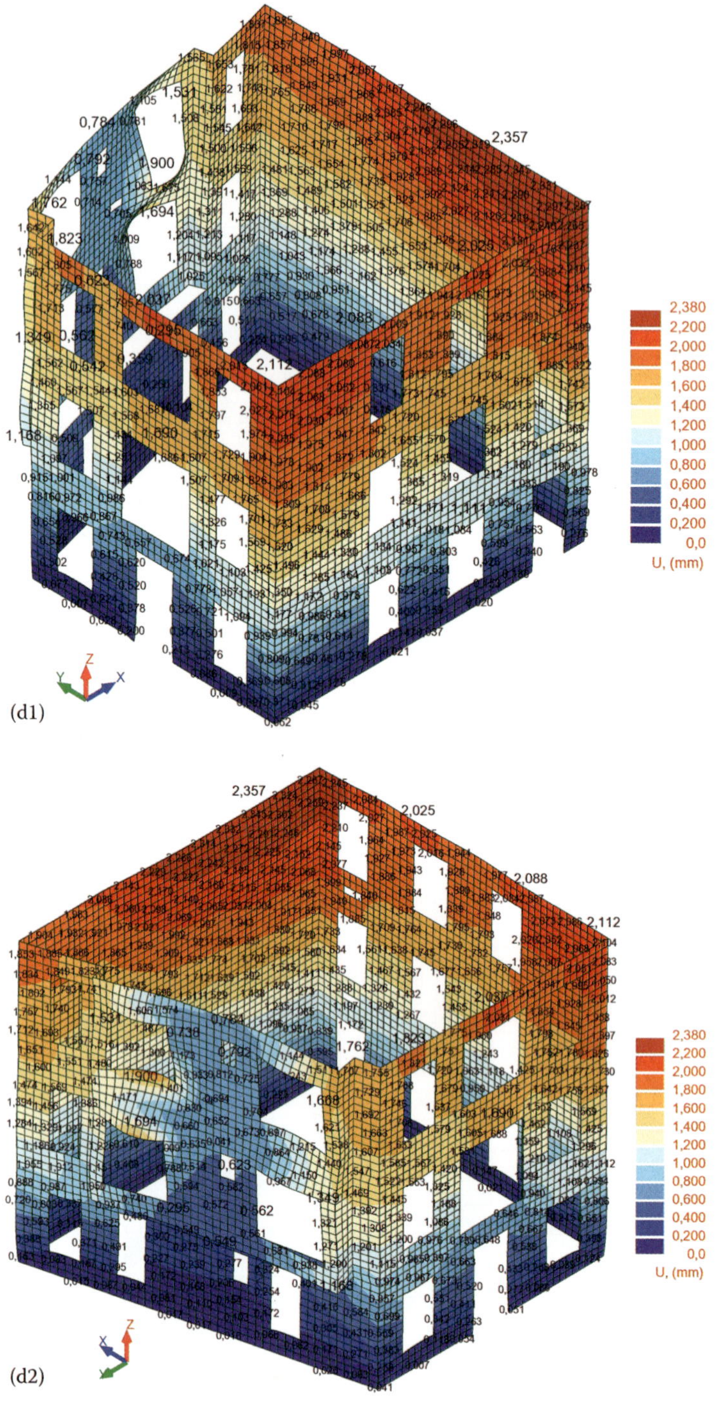

FIGURE 6.5 (Continued) (d) Global displacements for seismic action.

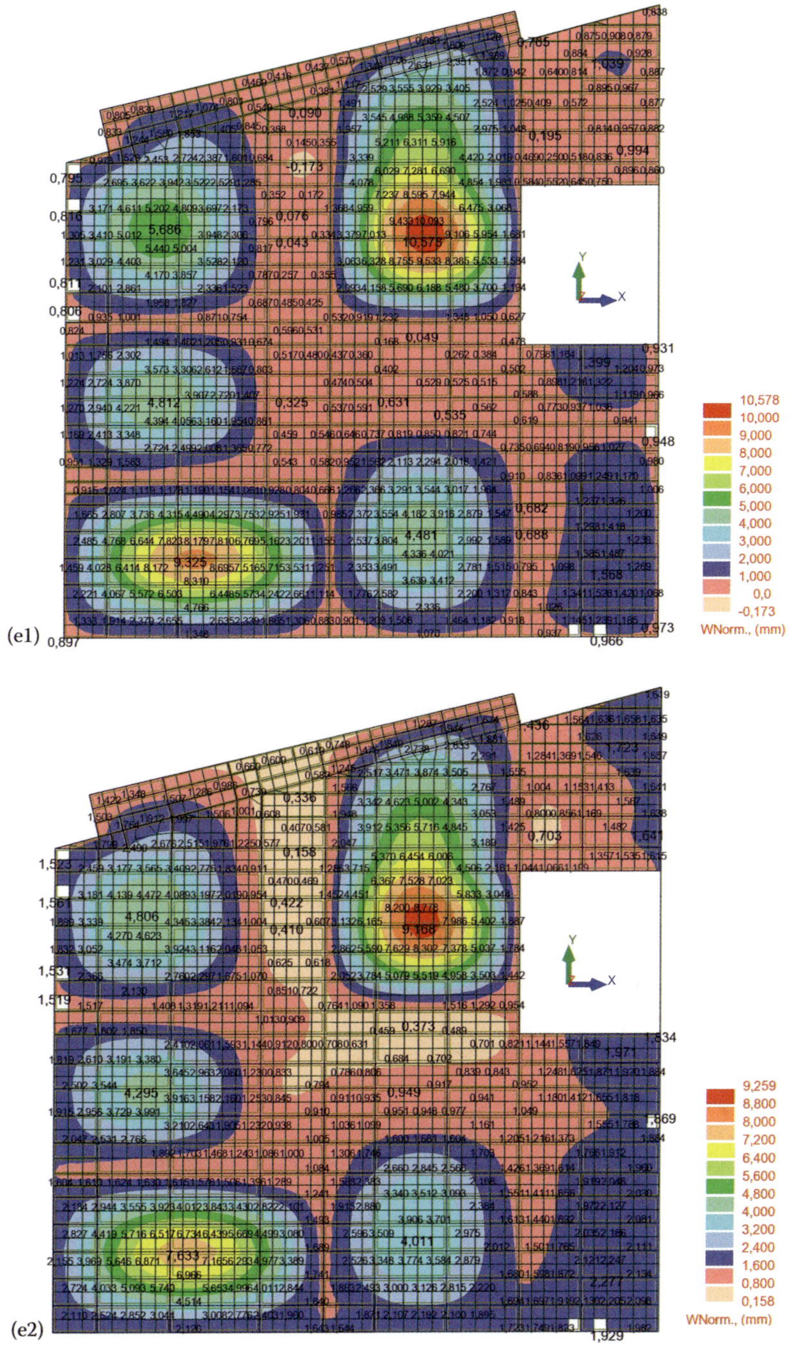

FIGURE 6.5 (Continued) (e) Deformation in wooden floors.

The case study presented here is a typical example of this. The building presents no visible damage in structural walls, coatings or roofs. Small maintenance works have been performed in the last years. A 3D model was developed and a structural analysis performed to identify the deformations and stress levels in the masonry and the *tabique* walls, wooden floors and roof structure. Furthermore, after the surveying, the material behaviour was assumed to be linear and was shown to be suitable. Nevertheless, more advanced methods exist and should be put into practice when possible, especially if the structures present visible damage. The material properties were set based on experimental studies conducted by the authors and co-workers. When these were absent, the properties were defined based on past research studies. The stress level, deformations and mode shapes have shown to be compatible with the current condition of the building.

Further research is necessary to improve the developed 3D model and thus to achieve results that are more accurate. Mesh refining and advanced analysis methods in areas with joints and section variations should be used to identify stress levels. The experimental studies conducted by the authors and co-workers in the historical old town of Leiria should be taken into account to increase the size of the database with results from tests on masonry and *tabique* walls.

REFERENCES

Asteris, P. G., V. Sarhosis, A. Mohebkhah, V. Plevris, L. Papaloizou, P. Komodromos, and J. V. Lemos. 2015. Numerical modeling of historic masonry structures. In *Handbook of Research on Seismic Assessment and Rehabilitation of Historic Structures*. ed. Panagiotis G. Asteris and Vagelis Plevris, 213–256. IGI Global.

Binda, A. S., A. Saisi and C. Tiraboschi. 2000. Investigation procedures for the diagnosis of historic masonries. *Construction and Building Materials* 14: 199–233.

EN 1015-3 1999. Methods of test for mortar for masonry. Part 3: Determination of consistence of fresh mortar (by flow table). European Committee for Standardization (CEN).

EN 1015-6 1998. Methods of test for mortar for masonry. Part 6: Determination of bulk density of fresh mortar. European Committee for Standardization (CEN).

EN 1015-7 1998. Methods of test for mortar for masonry. Part 7: Determination of air content of fresh mortar. European Committee for Standardization (CEN).

EN 1998-1 2004. Eurocode 8: Design of structures for earthquake resistance, Part 3: Assessment and retrofitting of buildings. European Committee for Standardization (CEN).

EN 1998-3 2005. Eurocode 8: Design of structures for earthquake resistance, Part 1: General rules, seismic actions and rules for buildings. European Committee for Standardization (CEN).

EN 338 2008. Structural timber: Strength classes. European Committee for Standardization (CEN).

ICOMOS 2003. Recommendations for the Analysis, Conservation and Structural Restoration of Architectural Heritage. ISCARSAH, International Council on Monuments and Sites.

Lourenço, P. B. 2002. Computations of historical masonry constructions. *Progress in Structural Engineering and Materials* 4 (3): 301–319.

Lourenço, P. B., J. G. Rots, and J. Blaauwendraad. 1995. Two approaches for the analysis of masonry structures: Micro and macro-modeling. *HERON* 40 (4): 313–340.

Pinheiro, P. 2015. Structural characterization of masonry walls from the Historical Center of Leiria. MSc diss. Polytechnic Institute of Leiria (in Portuguese).
Roca, P., M. Cervera, G. Gariup, and L. Pela. 2010. Structural analysis of masonry historical constructions. Classical and advanced approaches. *Archives of Computational Methods in Engineering* 17: 299–325.
Santos, S. P. 2003. Structural rehabilitation of historic buildings. *40th Meeting of Commission CIB W023*. Padova, Italy.

7 Characterisation and Rehabilitation of One of the Historical Buildings of the Polotsk Campus of the Polotsk State University

Yahor Lazouski, Aliaksandr Paprauka, Dzmitry Lazouski and Aliaksandr Bakatovich

CONTENTS

7.1 INTRODUCTION

The architectural ensemble of the Polotsk Jesuit Collegium is an eighteenth-century monument of late Baroque architecture. One of its masonry buildings has been almost totally damaged over its lifetime. The authors present their practical experience in the field of surveying, characterisation and rehabilitation of this building (one of the PSU Polotsk campus houses, which is now situated in the ensemble of buildings of the ancient Jesuit Collegium). A short historical review of the building and Collegium, the nondestructive methods for the determination of the properties of the structural materials, the decisions on how to restore the carrying capacity and the strengthening methods that were applied to this restored building are presented in this chapter. Special attention is given to the authors' improved method of increasing the rigidity of masonry buildings.

7.2 A SHORT HISTORY OF THE POLOTSK JESUIT COLLEGIUM

The building of the Jesuit Collegium is a Baroque architecture monument of the eighteenth century. The Collegium was inaugurated in 1580 by decree of the Polish King Stefan Batory. The first Rector of the Polotsk Jesuit College was the well-known preacher Peter Skarga. During the eighteenth and nineteenth centuries, the Collegium was one of the richest in the territory of modern Belarus. Horse stables, a coach, a bakery, a smokehouse, a brewery, workshops and a cloth factory were built there between 1773 and 1780. The Collegium also contained a chemist's, a hospice and a music school. Teachers of different disciplines were being prepared to teach in other Jesuit schools.

By decree of Emperor Alexander I, the Polotsk Jesuit Collegium was converted into an academy with all the rights and privileges of a university. Jesuit schools during the Russian Empire were under the jurisdiction of the Polotsk Jesuit Academy. The grand opening of the Polotsk Jesuit Academy took place on 10 June 1812, and the Belarusian Governor-General Duke Alexander of Württemberg attended in person.

The Polotsk Jesuit Academy published the first scientific journal about modern Belarus (it was not about modern Belarus, but it was the first on the territory where Belarus is situated now), *Mesechnik Polotsky*. During 1818 and 1820, there were 14 publications, which contained 85 articles.

The Polotsk Jesuit Academy held a rich library (more than 50,000 books). At the Jesuit College and the Academy, there was a printing house, which released about 500 titles from 1787 to 1820 (mostly in the Polish language).

The huge museum, which was located in a separate three-storey building, was an outstanding attraction of the Polotsk Academy. There was a famous mechanical head, made by Gabriel Gruber. It had movable eyes and could answer any questions asked by visitors using no matter what European language. The museum also had a large collection of weapons, paintings, prints, geological and zoological exhibits, and various mechanisms and devices.

The large astronomical hall had a unique eight-foot telescope. The physical and mechanical rooms with sparse equipment, which had been built by the teachers of the Academy, the chemical laboratory and the mineralogical room were the pride of the Academy.

In the nineteenth century, the Polotsk Jesuit Academy was the largest educational centre in modern Belarus. There was a famous confrontation between the Polotsk Jesuit Academy and the Vilnius University, of which the latter had adhered to a totalitarian ideology followed by the French Enlightenment. The Polotsk Academy, instead of teaching Polish, taught Russian, which had taken precedence and had become more consistently incorporated into the educational process. The Polotsk Jesuits contributed to the strengthening of the regional identity of the Belarusian local nobility, as opposed to the Polish policy of Vilnius University. Prominent figures such as Barschevsky Jan, Tadeusz Dzhovsky, Gabriel Gruber, Francis Kara, Adam Kripski and Konstantin Tyszkiewicz were among the graduates of the Collegium and the Academy.

In 1814, when Pius VII reinstated the Order, the missionary work of the Jesuits in the Russian Empire began to cause a sense of unease among the authorities. The Jesuits were banished from St Petersburg to Polotsk in 1815. In 1820, after the death of General Order Thaddeus Brzozowski, the Jesuits, who did not wish to abdicate their vows, were asked to leave the Russian Empire, and the Polotsk Academy was abolished.

After the expulsion of the Jesuits from the Russian Empire, a significant part of the Polotsk Academy's library was ransacked. What remained was transferred to the Polotsk Piarists. In later stages, the preserved books were then distributed among various Russian educational institutions. The printing equipment that belonged to the Polotsk Academy was sent to Kiev in 1833.

From 1822 to 1830, the buildings of the former Polotsk Jesuit Academy housed the School of Public Relations at the highest level. In the years 1830–1835, the Academy building was reconstructed, and a new building was built. From that time on, the Polotsk Cadet Corps were located on those premises.

In Soviet times, a military hospital was located on the Academy's land. Since 2005, the History and Philology Faculties and the Faculty of Information Technologies of the Polotsk State University have taken up residence in the preserved building.

7.3 THE BUILDING OF HOUSING 'ZH' OF THE POLOTSK CAMPUS OF THE POLOTSK STATE UNIVERSITY: THE NONDESTRUCTIVE METHODS AND THE RESULTS OF THE TECHNICAL SURVEY OF THE BUILDING

Nowadays, housing 'Zh' is one of the buildings found on the Polotsk campus in the Polotsk state University. Housing 'Zh' was originally built to be a Collegium music school. This is one of the last Jesuit buildings of the eighteenth century. This building was built according to the methods of a very simplified scheme. Its walls are thinner than the walls of any of the other buildings of the complex, and the upper floors were overlapped using wooden beams instead of masonry arches. The floors in most of the rooms, including the dining room on the first floor, were made using the 'warm' design (warm air from the furnace under the floor covering).

Nowadays, the main part of housing 'Zh' is a two-storey rectangular building without a cellar, the dimensions of which in the edged axes are 14.6×10.75 m. There are also two-storey extensions of the main building near the transverse axes '2' and '5' with plan dimensions 4.55×5.35 m and 3.00×2.85 m, respectively. The plans of the first and second floors of the building are shown in Figure 7.1.

The external and internal load-bearing walls of the building are made with the thicknesses of 700, 800 and 900 mm. These walls are made of red clay bricks, produced using technology usually used in plastic formation. The strength parameters of the bricks are determined using the nondestructive shock-pulse method (using

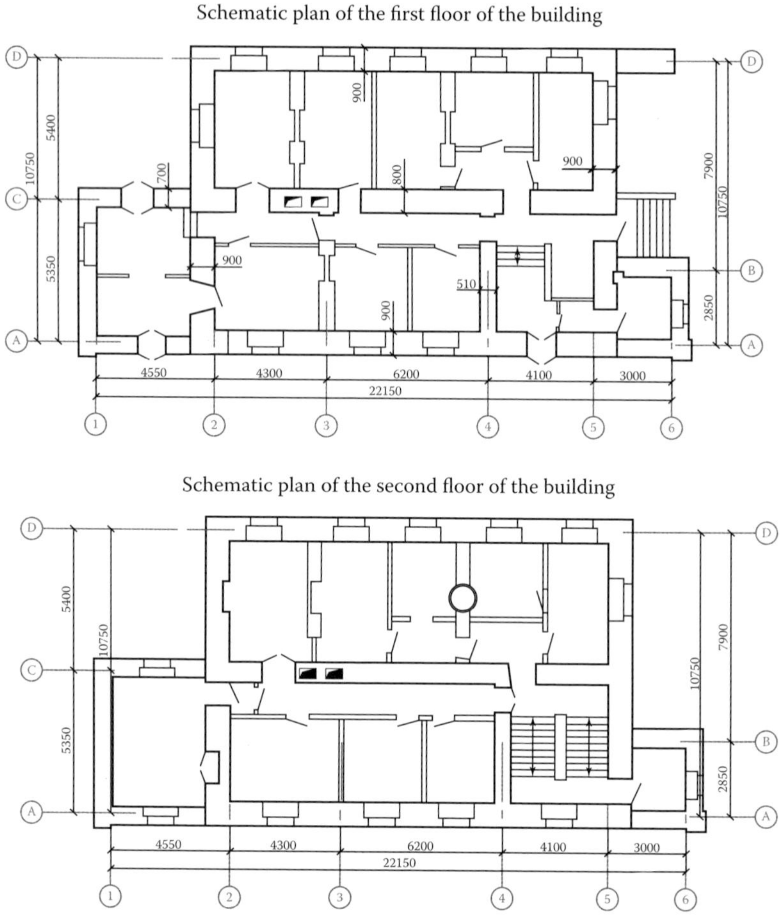

FIGURE 7.1 Schematic plan of the first and second floors of the building.

such types of devices as the Schmidt hammer and special instruments for compressive strength control [such as IPS-MG4.03]). They were also tested until the breaking point of a certain number of bricks, which were extracted from the original masonry wall. The minimum strength of the bricks during the compression test was 11.2 MPa (GOST 8462-85).

The lime-sand mortar was also tested to obtain certain parameters. The compressive strength of the mortar was obtained by testing five sample cubes with an edge length of 40 mm, made of two or three mortar plates taken from the horizontal masonry joints and united by using a gypsum solution (Recommendations 1987). The compression strength of the mortar obtained using this test was higher than 5 MPa.

To search for metallic elements in the masonry walls, a Hilti Ferroscan PS200 was used. During this operation, further steel joints were found in the internal and external walls, and this led to a better and even more precise development analysis model of the building.

The building of housing 'Zh' has wooden beams in the ceilings. The hip roof of the building was made with an external unorganised water drain. The rafter roof system of the building was composed of wooden trusses (principals) of a triangular shape.

The technical survey of the masonry walls of the building revealed that they had suffered damage. That damage had adversely affected the strength and the durability of the walls. There were numerous cases of dangerous structural damage, such as the soaking of the brickwork in places where there was an absence of organised water drainage from the roof, which in turn led to the breaking of the plaster coating. Cracks with widths up to 15 mm were also revealed during the survey of the building. Furthermore, hidden damage was found in the masonry walls during the reparation and survey work. This ran along the entire length of axes 'C' and '4' in the form of

- Numerous holes in the places of the chimneys with structural destruction shown in the brickwork
- Through cracks crossing more than 10 rows of masonry
- Irregular yielding of the walls with subsequent destruction of lintels over the doorway
- Destruction plots of the brick masonry above windows and doorways

The location scheme of the characteristic damage of the exterior walls of the building façades is shown in Figure 7.2.

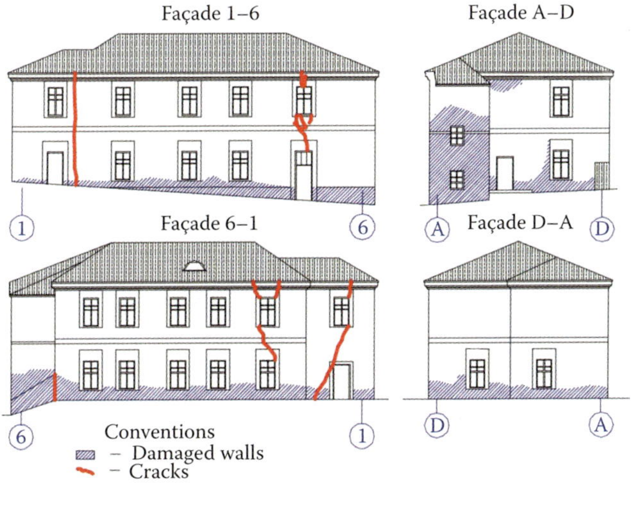

FIGURE 7.2 Location scheme of the characteristic damage of the exterior walls of the building façades.

The crack width in external and internal masonry walls increased during the observation period of one year.

The rot due to the bad condition of the roof covering seriously damaged some of the wooden trusses (principals) of the rafter roof system. The building's floor and windows were completely destroyed. The first-floor coverings had suffered damage.

The main reason for the destruction of the walls was nonuniform yielding of the foundations. The technical survey of the soils of the building foundation by ground-penetrating radar sensing revealed a difference of the density characteristics of the foundation soils. This is the main reason for the nonuniform yielding of the foundation of the building walls, which was caused by the violation of the project (normal) operating conditions:

- Certain changes in the wall stiffness and loads on foundation sites made to the additional window and door openings during the dismantling and restructuring of the parts of the building
- Increased loads on the ceilings in relation to the original design
- Changes in the physical and mechanical characteristics of the soil of the foundation due to poor drainage of water and even the wastewater systems
- Destruction of the walking perimeter
- Trees growing near the foundation
- Changes in the ground level starting with a slope to the building
- Changes in temperature in the building – the lack of heating in the autumn and winter periods, and so on

There is no walking perimeter around the building. This promotes the penetration of rain and meltwater from the surface into the soils of the foundation, so reducing their physical and mechanical strength and deformation properties.

The appearance of the detected damage of the masonry walls is shown in Figure 7.3.

7.4 REPARATION AND RESTORATION WORKS ON HOUSING 'ZH' CARRIED OUT ACCORDING TO THE RESULTS OF THE TECHNICAL SURVEY

As a result of the technical survey, for the continued reliable operation on the housing 'Zh' building, a design was made for the reconstruction project of the building.

To restore the loading capacity of the masonry walls of the building, the following activities were performed:

1. The damaged sections of the masonry walls were rebuilt:
   - Along the axis 'C' within the second floor
   - Along the axis '4' to the full height of the wall
   - Along the axes 'A', 'D', '1' and '6' to the damaged areas.

FIGURE 7.3 Appearance of the detected damage of the masonry walls.

2. The damaged masonry lintels of the windows and the doors were strengthened. The strengthening procedures were accomplished by installing additional internal steel lintels, embedded into the wall with both edges to a depth of at least 150 mm.
3. New reinforced concrete ceilings were placed (Lazouski, 2013).
4. Due to the fact that the masonry building has a significant number of vertical cracks, including nonuniform foundation yielding, and the cracks will continue to increase over time, an output of the insufficient stiffness of the whole building was prepared. To increase the building's stiffness, steel pre-stressed flexible belts were set in place. Steel pre-stressed flexible belts are a system of horizontal ties, made of smooth steel reinforcement bars that are Ø32 mm, which created several closed contours around the walls' perimeter. To preserve the architectural appearance of the building, all strengthening elements were established in the pre-formed indents. The pre-stressed rods were anchored by welding their fixation at its ends to the rolled steel standard section elements, which were installed at the corners of the building. Belt tension to the value of 100–120 MPa was carried out

by using steel couplings on the right- and left-hand thread placed in the middle of the belt sections. This resulted in the compression of the whole building, which increased its stiffness and helped redistribute the load to the building's foundation more uniformly.

Experience shows that pre-stressing decreases over time due to various processes, such as masonry creeping, the temperature dropping, and so on. Bands become slack, and so the effect of compression is lost.

The ongoing processes of nonuniform foundation yielding have produced further disclosure of the cracks in the walls. The increasing of the cracks' width has increased the compression efforts. However, the longer the building or wall, the less effect of additional compression is found or seen. For example, additional crack opening with a width of 10 mm in buildings with a length of 40 m has led to a rise in stress in the belt rod, but by only 50 MPa.

A method (Lazouski et al. 2001) that resorts to pre-stressed belts with intermediate rigid anchors that are located along the belts' length was designed to improve the effectiveness of the method and increase the stiffness in the masonry building, which is located on the premises of the Polotsk State University. The belt ties were set to anchors through the welding or the bolting of their ends. The intermediate anchors allow working tension to be applied to the stonework in the places around them, so allowing the cross-sectional area of ties to be reduced with the distance from the crack. This reduces the unsupported length of the strand and improves the efficiency of the response of any additional width shown in the crack openings (the additional crack opening width of 10 mm in the building with a length of 40 m with three intermediate rigid anchors led to an increase of the stress in the belt rod of about 200 MPa, with four intermediate rigid anchors, to an increase of the stress in the belt to 250 MPa, etc.).

As seen from the example, the proposed enhancement allows the use of steel rods with higher strength to increase the stiffness of the masonry buildings. Reducing the length of the strands also eases the process of pre-stressing, which allowed the use of thermo-mechanical tensioning in a more effective way.

The building of housing 'Zh' has been strengthened by three steel flexible pre-stressed belts with intermediate anchors to ensure the necessary level of stiffness to prevent the increase in the width of the cracks and enable further reliable exploitation, as is shown in the schemes (Figure 7.4).

5. After the operations to increase stiffness, the damaged wooden trusses (the principal ones) of the rafter roof system were changed. After that, the new roof covering was made.
6. The exfoliating plaster was removed from the walls, followed by the reduction of the protective coating and the implementation of operations to eliminate water-soaked walls (making the new drainage system along the perimeter of the building).
7. New doors and windows were installed.
8 On the first floor, the floor coverings were restored.

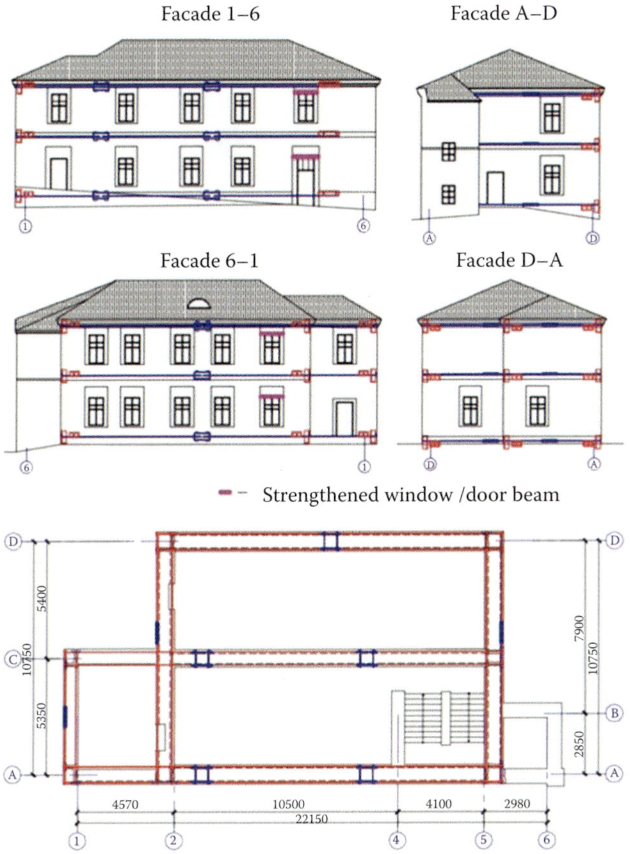

FIGURE 7.4 Schemes for increasing the building stiffness and strengthening the walls.

9. Once it was certain that the necessary rigidity of the building had been achieved, the interior and exterior finishing was performed.

7.5 CONCLUSION

The development of new nondestructive methods for the testing of building structures, along with the development of strengthening methods and techniques, allows existing historical buildings to be recovered and reconstructed, thus extending their life cycle. Thus, after the reconstruction and restoration of the housing 'Zh' building, it is now being used for housing on the Polotsk campus of the Polotsk State University, as it was in the past, when it was used as a part of the educational establishment. Therefore, the new developments are helping to preserve the historical heritage. A general view of the building of housing 'Zh' before and after the reconstruction is shown in Figure 7.5.

FIGURE 7.5 General view of the building before and after the reconstruction.

REFERENCES

GOST 8462-85. 2001. Wall materials. Methods for determination of ultimate compressive and bending strength IPK Publishing House of Standards.

Lazouski, D., N. Maseenok, and E. Matskevich. 2001. The reasons for non-uniform sediment of the foundations and strengthening of the walls of stone buildings. *Architecture and Construction* 6: 47–49.

Lazouski, Y. 2013. Experimental and theoretical studies of the strength of the bending reinforced concrete elements, strengthened with additional prestressed transverse reinforcement. *Herald of Polotsk State University. Series F. Civil Engineering. Applied Sciences* 16: 51–60.

Recommendations for survey and quality assessment, of constructed and operated structures, using non-destructive methods. 1987. NIIZhB, Stroyizdat.

8 Applications of GPR in the Study of Cultural Heritage

Examples of the Ability of the Method in Assessing Different Features

Vega Pérez-Gracia, Sonia Santos-Assunçao, Oriel Caselles and Ramón Gonzalez-Drigo

CONTENTS

8.1 INTRODUCTION

Cultural heritage is part of the legacy of civilisations. The general term includes all the different expressions of past generations, comprising customs, language and artistic expressions, among others. Some aspects are represented by the built vestiges: buildings, bridges, statues, remains of cities and villages, or roads. These structures are the tangible representation of the values, technology and organisation of past cultures and must be studied and preserved to understand our past and to pass the legacy on to the next generations. The preservation of the built environment involves in most cases a deep study of the structures. Two of the main reasons for that study are to obtain meaningful knowledge of the structure and to detect damage that needs to be repaired. Without those two aspects, the design of the maintenance and possible restoration can be impossible or not accurate enough. The analysis of the structures belonging to the built cultural heritage environment usually requires nondestructive tests (NDT), which supply the data needed to evaluate the construction and to decide on further actions to preserve and maintain it. However, NDT are usually based on indirect measurements. One of the most popular methodologies in the NDT of cultural heritage is ground-penetrating radar (GPR). It is a fast and effective method to obtain information about the state of masonry structures, although it presents certain inherent limitations to any indirect study. In this chapter, two examples of the assessment in cultural heritage buildings are presented. The first case study is based on the analysis of masonry, detecting damage and embedded targets. This study is completed with seismic tomography, which is also an NDT technique, and endoscopy (a slightly destructive technique). The second case is focused on the evaluation of wooden beams. In many old buildings, wooden elements are combined with masonry, presenting problems due to humidity or woodworm.

8.2 CASE 1: ASSESSMENT OF MASONRY COLUMNS

8.2.1 Introduction

The group of buildings of the old 'Hospital de la SantCreu I Sant Pau' is one of the most representative Modernist buildings in Barcelona. That architectural ensemble has been a UNESCO World Heritage Site since 1997, and building began in 1902 for a hospital. The architect Luís Domènech i Montaner designed a project with 48 pavilions (Domènech i Montaner, 1916), even though only 13 were built in the first phase (1902–1913) and 6 were added in the second phase in 1920. Those buildings replaced the oldest hospital created in 1401 through the union of the different hospitals that existed in Barcelona.

In April 2004, the main dome of the gynaecological ward collapsed on a patient waiting room. This dome was being restored at the time of the accident. The deterioration of ageing materials and the vertical fissures on distinct metallic axes were considered the main causes. On the other hand, the buildings were situated in an area characterised by shallow geological heterogeneities.

In 2009, the medical services were completely relocated to a new building near the zone, and the Modernist buildings were intensively assessed.

In most parts of those buildings, columns are essential elements to distribute the roof loads and to diminish the weight on the main walls. A large number of them are distributed in different zones (Figure 8.1), their appearance being quite similar: varnished clay bricks. However, the visual inspection of the columns in different pavilions denotes the existence of damage: cracks, humidity, change of colour and loss of the varnish layer (Figure 8.1), even though, in some cases, the columns were repaired with rings years ago.

There were three main objectives of the nondestructive assessment of the buildings: (1) detecting damage not observed during the visual inspection, (2) defining the extent of the damage observed in the different elements and (3) determining

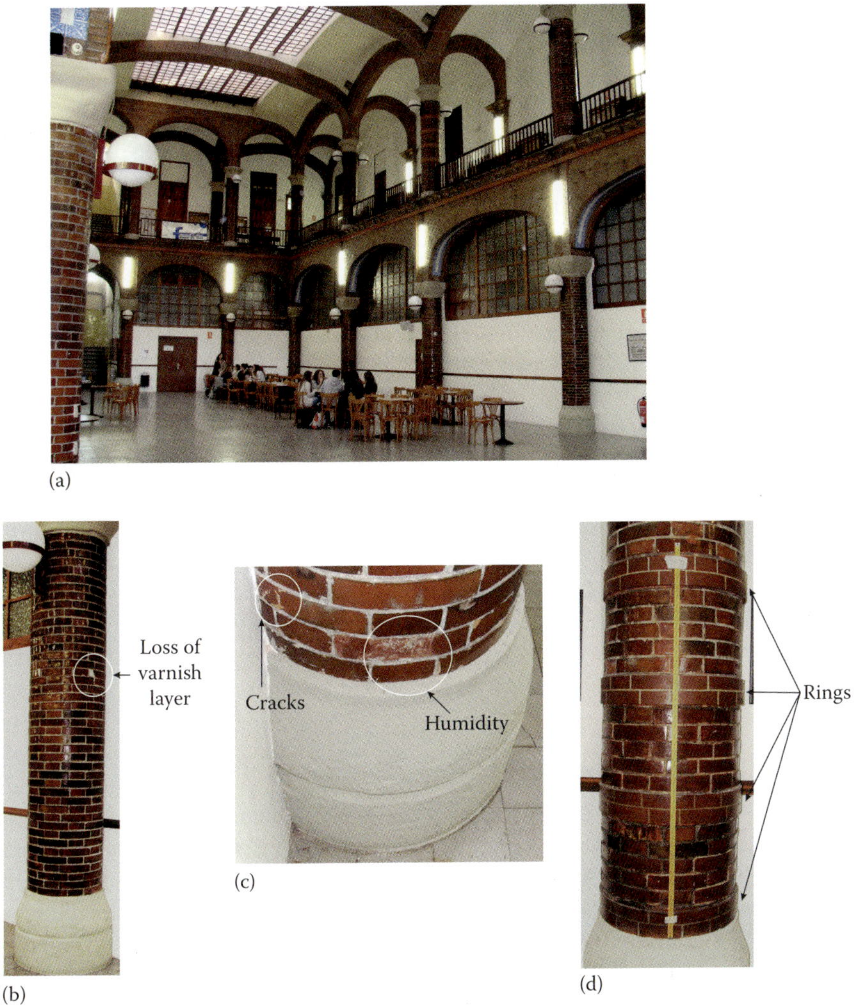

FIGURE 8.1 (a) Distribution of the columns in one of the pavilions. (b, c) Damage detected in the visual inspection. (d) Old restoration of one column using rings.

the inner structure of the supporting elements. The three aspects were necessary to assess the overall damage of the structures and also to define accurate models with which to simulate the behaviour of the structure under different loads and vibrations.

As an initial analysis, three methods were applied in the evaluation: GPR, seismic tomography and endoscopy. Both GPR and seismic tomography are nondestructive tests providing data that can be interpreted in a general overview. Endoscopy is used to visualise in situ the interior of the structure at selected points, being considered a slightly destructive method.

GPR is considered a useful tool to define the geometry of the internal structures and to map damage in masonry (e.g. McCann and Forde 2001, Binda et al. 2005, Diamanti et al. 2008, Pérez-Gracia et al. 2011), this technique being widely applied in cultural heritage (e.g. Leucci et al. 2007, González-Drigo et al. 2008, Solla et al. 2011) and to detect embedded targets in masonry, such as stones, wood and steel (e.g. Binda et al. 1998, Santos-Assunçao 2014). In those cases, the masonry materials present enough electromagnetic contrast compared with the embedded targets to allow their detection. However, in many cases, the use of a single technique is not enough to obtain accurate results. It is usual to combine different NDT methodologies to determine with maximum accuracy the state and inner constitution of the constructive elements (e.g. Nuzzo et al. 2009, Pérez-Gracia et al. 2013, Martinho and Dionísio 2014).

Combining methodologies is always recommended in the case of complex structures or when information about materials is needed. In the first case, data obtained from different methodologies can clarify the uncertainty inherent in all indirect methods. The images from complex media are usually confused and obscure. The integration of different methods and parameters allows anomalies that are not clearly observed to be highlighted and identified. Besides, GPR images denote the existence of anomalous targets or discontinuities, but it is difficult to determine detailed aspects related to the physical properties of materials. The combination of GPR with other techniques provides, in many cases, data about those physical properties. That is the case when it is combined with seismic or acoustic data that allow the elastic wave velocity to be known, related to the materials and their state of conservation, or combined with resistivity surveys, allowing maps of resistivity that can be used to distinguish between voids and material targets, for example.

8.2.2 Methodology

The GPR survey with a 1.6 GHz centre frequency antenna was focused on the columns and the elements connecting the columns. The position of the antenna along each element was defined with a survey wheel.

Columns are wrapped with horizontally disposed ceramic bricks. Depending on the age of the building, columns may have internal metallic structures to improve their strength. At the end of the nineteenth century, the reinforcement of masonry elements was done using two-by-two rings in a U or L shape, vertically distributed along the column. However, in other cases, no metallic reinforcement was used in the columns. In these cases, the bricks were bulky, solid and compact. The lack of constructive information does not allow us to know whether this type of internal reinforcement was used in some of the columns, even though some of the observed damage on the surface of the columns could correspond to corrosion effects. Oxidation, a chemical process, is considered a corrosion process, which often occurs with ceramic bricks as a consequence of their mineralogical content. The bricks composed of clay elements (aluminium silicates) are vulnerable to the combination of water and oxygen, increasing the corrosion. As a consequence, the oxygen appears on the surface and not in the brick junctions. In the pavilion's studied columns, the surface is fully protected by varnish, which minimises the described pathology, whereas the bricks' unions (not protected with the varnish) allow water intrusion. Hence, the affected area is a point that expands in a circular shape. In many cases, in the inner part of the column, tubes were used for the drainage of water. Combined with atmospheric humidity, the effect on the bricks and mortar produced significant damage.

Radar data was acquired along horizontal radar lines (at the perimeter of the column) spaced 4 cm and along vertical radar lines (at the shank of the column) spaced 5 cm apart (Figure 8.2). A time window of 8 ns was used in all the radargrams, the spatial sampling being 0.2 cm. Each of the radargrams was processed first with ReflexW software (Sandmeier 2003) and second with the GPRSlice software (Goodman 2004, Goodman et al. 2008) to obtain 2D time slices and 3D radar images of the inside of the columns. The processing applied to each B-scan was a sequence composed of the search for zero time, manual gain, subtract DC-shift, dewow and bandpass filter with cutoff frequencies of 500 and 2200 MHz. The second part of the processing at each B-scan obtained along the perimeter of the column was to wrap the radargrams to obtain a circular shape. After that, the horizontal and vertical scans were interpolated to obtain 3D radar images and time slices by using the square of the wave amplitudes. The last step was obtaining the isosurface rendering imaging using a cutoff amplitude of 75% (Santos-Assunçao et al. 2014).

Seismic tomography was based on the transmitted wave amplitude using only the first arrivals. Seismic data was obtained along the diameter of the columns and, as a result, defined 2D images that indicate changes in the wave velocity. Geophones were placed at 13 points along the diameter of the column (see Figure 8.2), and the seismic source was an instrumented hammer.

Seismic tomography is obtained by dividing the assessed surface into cells, the results being the sum of the values in each of those cells. The trajectories in homogeneous media are defined in Figure 8.3. In nonhomogeneous media, seismic waves are refracted in the discontinuities between elements characterised by different velocities. In the computational process, seismic velocities are associated with the different cells, and refraction is caused at adjacent cells with different velocities. An iterative process is continued until the convergence of the solution, taking into

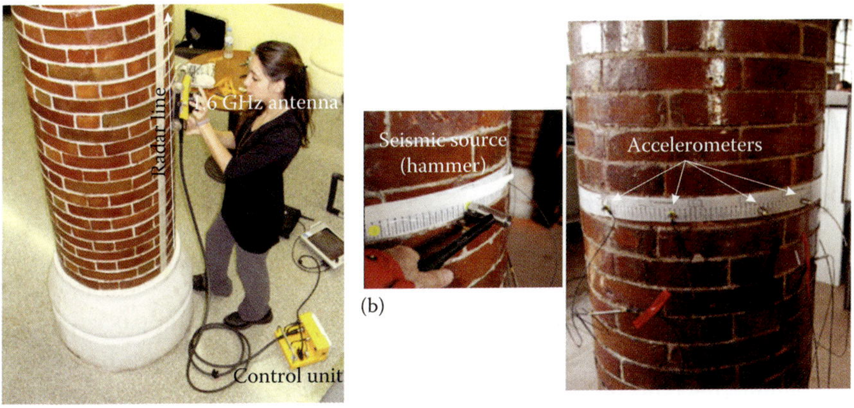

(a) (b)

FIGURE 8.2 (a) Radar data acquisition. (b) Seismic data acquisition.

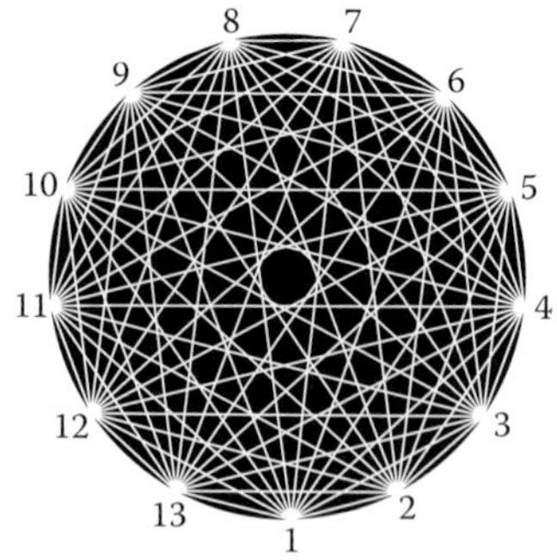

FIGURE 8.3 Trajectories of the seismic waves between the 13 measurement points in a homogeneous medium.

account the rays' curvature as a consequence of refractions (Michelena et al. 1993, Sandmeier 2003).

In the case of nonhomogeneous media, the seismic wave is refracted because of changes in the wave velocity associated with adjacent cells, and equations are solved in an iterative computational process until convergence of the solution. The computational process, the simultaneous interactive reconstruction technique (SIRT), contemplates ray curvature as a consequence of internal refractions (Michelena et al. 1993; Sandmeier 2003). The characteristics of each cell are defined in the case that at least one ray path crosses the cell.

8.2.3 Results

Vertical GPR profiles denoted a large number of anomalies. In all of the B-scans, small hyperbolic anomalies appeared near the first arrivals, corresponding to the 1 cm junctions between the bricks in the superficial layer (see Figure 8.4). Radar data show a constant 6 cm distance between these hyperbolas, which corresponds to the 6 cm wide bricks. At a depth of about 3 ns, a constant anomaly could correspond to a contact between bricks and other materials or further layers of bricks. At about 5 ns, a strong reflection appears irregularly along all the radargrams. This reflection disappears in some zones but presents a high amplitude in other parts of the column.

Horizontal profiles, after wrapping the B-scans, reveal a heterogeneous medium (Figure 8.5).

The correspondence between the wrapped images and the raw B-scans is shown in Figure 8.6. The strong anomaly observed at about 5 ns corresponds to the anomaly labelled B in Figure 8.5.

The analysis of the different columns highlighted the heterogeneous medium, which is not composed of regular and parallel bricks. On the other hand, in many of the columns, the reflection on the internal target appeared more clearly in some sections, while in others the reflections were strongly diminished. In addition, in some columns, this anomaly was not recorded. 3D images were built to better understand the anomalies (Figure 8.7).

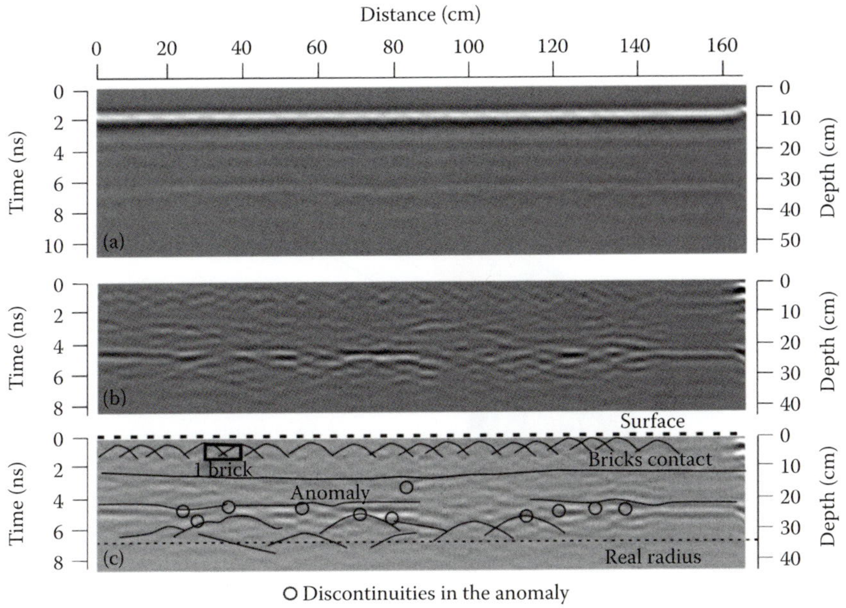

FIGURE 8.4 Radar data along one column (vertical profile). (a) Raw data. (b) Processed data. (c) Possible radar data interpretation.

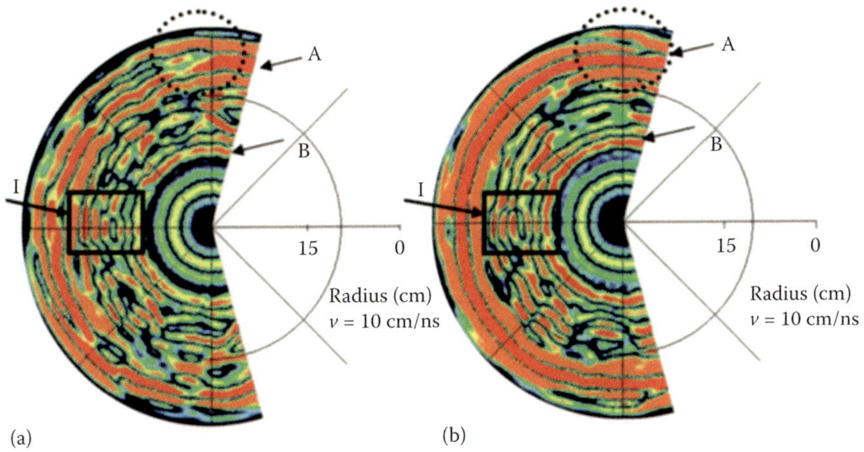

FIGURE 8.5 (a) and (b) Two wrapped B-scans from the same column. The distance between both sections is 13 cm. Rectangular (I) and circular (A) marks denote anomalies that could correspond to the same target, visible in both profiles. The label B indicates the existence of a regular anomaly that is observed in all profiles at the same two-way travel time.

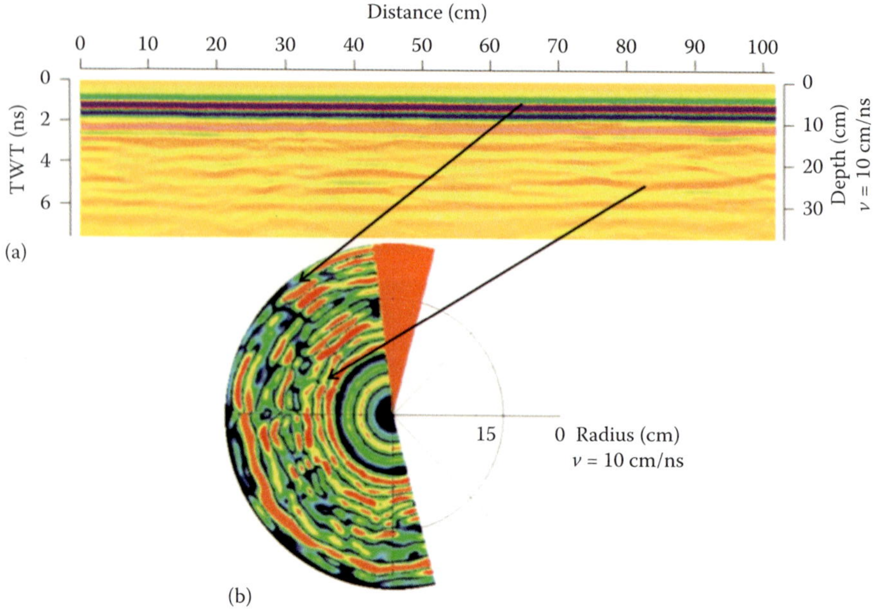

FIGURE 8.6 (a) Raw B-scan in rectangular coordinates and (b) the same B-scan after wrapping to the real shape of the column section. The arrows allow the correspondence of anomalies to be identified.

To solve the uncertainties, seismic tomography (Santos-Assunçao 2014, Santos-Assunçao et al. 2014) was compared with some of the GPR sections. Figure 8.8 shows the image of one of the tomographic sections in which the lower wave velocities were associated with fissures or damage. The images denote dramatically heterogeneous materials and obscure the central area corresponding to the zone inside the anomaly in the radargrams. Even though the seismic tomography determines the existence of many contacts and discontinuities, it was not possible to define the shape of the bricks inside the column.

Endoscopy was also used to solve the uncertainties of the highly heterogeneous media. This method consists of introducing a borescope into a tiny drilled opening hole (habitually with a diameter of less than 12 mm) to examine the interior of the structures. The borescope is a bendable tube with a light source and an eye-piece. Images obtained in some selected columns validate the existence of diverse heterogeneous materials and most likely a metallic element (Santos-Assunçao 2014). In some situations, cracks appear on the images associated with bricks, highlighting some of the pathologies affecting the interior of the column (Figure 8.9).

Diverse materials are distinguished. Corroded metallic elements and eroded bricks indicate inner damage. Some interfaces are not well defined due to the high degree of damage.

Finally, one of the columns was subjected to a destructive load test. This last essay allowed the images from the NDT methods to be compared with the results (Santos-Assunçao et al. 2014). The images of the interior of the column show the

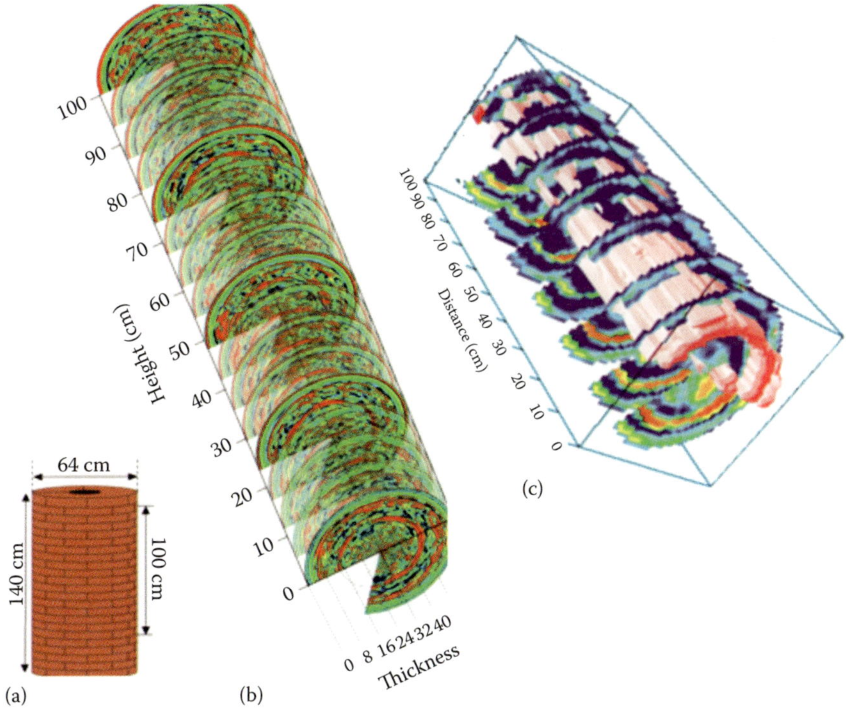

FIGURE 8.7 3D images of one of the columns. (a) Scheme of the column. (b) Transparent column including all the B-scans. (c) Render isosurfaces.

irregular arrangement of bricks, corresponding to the heterogeneous anomalies in the B-scans (Figure 8.10). The existence of an iron drainage tube is remarkable. Part of this tube is highly corroded and has disappeared. The 3D reconstruction of the internal anomaly corresponds to this damaged tube.

8.3 CASE 2: STUDY OF WOODEN BEAMS

8.3.1 Introduction

Wooden structural members are usual in old and heritage buildings. In Barcelona, these structures can be found in two areas: in the Eixample district, built during the nineteenth and twentieth centuries, and in downtown Barcelona. In the first zone, the typical buildings combine masonry with concrete and wooden beams. In the second zone, numerous houses built during the seventeenth century or before exist, wood comprising most of the supporting material, combined in most cases with masonry.

The age of these kinds of structures requires rigorous maintenance, because the wood can be extremely damaged. Fungi, woodworm and damp used to be the biggest enemies of these structures. As most of the beams, and in some cases, columns, are wooden, it is important to assess their state of conservation to preserve and maintain the building. The first problem with direct inspection is that direct access to the

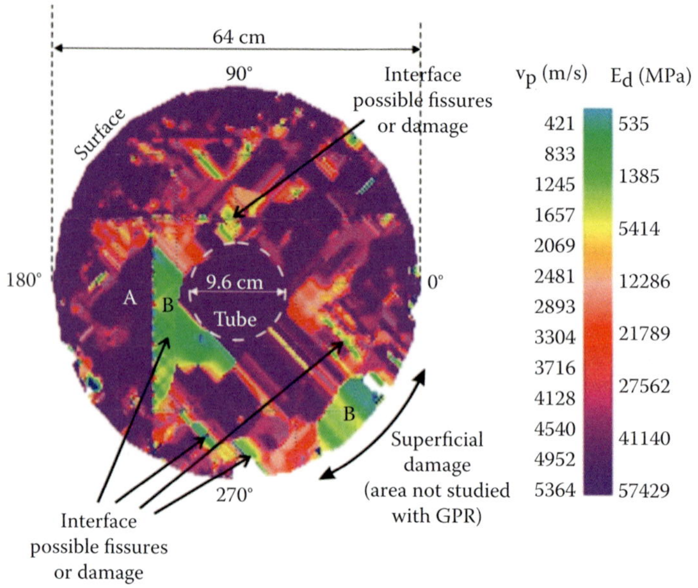

FIGURE 8.8 Seismic tomography of the column. The central zone is associated with a tube and corresponds to the area inside the central anomaly of the wrapped B-scans. Label A marks a zone characterised by high velocity and label B, a zone with low velocity. B areas could be related to higher damage, because they are due to higher air content. A clear interface between A and B is detected, and possible fissures are indicated in several zones.

wooden beams is not always possible. In addition, destructive testing is only recommended in the case of a high probability of damage. This background requires the use of NDT.

The study presented in this chapter is the assessment of the wooden beams in a building in downtown Barcelona. The study was carried out because some highly damaged beams were discovered during maintenance tasks on the floor tiles (Pérez-Gracia et al. 2014). Figure 8.11 shows some details of the beams in connection with the load-bearing wall.

The cause of the damage in the heads of the beams was the moisture of the masonry wall that supported the beams. The connection between the beams and the wall is made by means of masonry boxes, the wood being in direct contact with the bricks (Figure 8.12).

8.3.2 Methodology

A 900 MHz centre frequency antenna was used to assess the beams. The access was only possible from the floors, due to the rooms being small. Therefore, only a common offset survey was possible. The time window was 15 ns, and the data was acquired with 32 scans/s.

The assessment process was carried out in four stages, the first of which was a visual inspection of the rooms, paying special attention to the zones where the beams

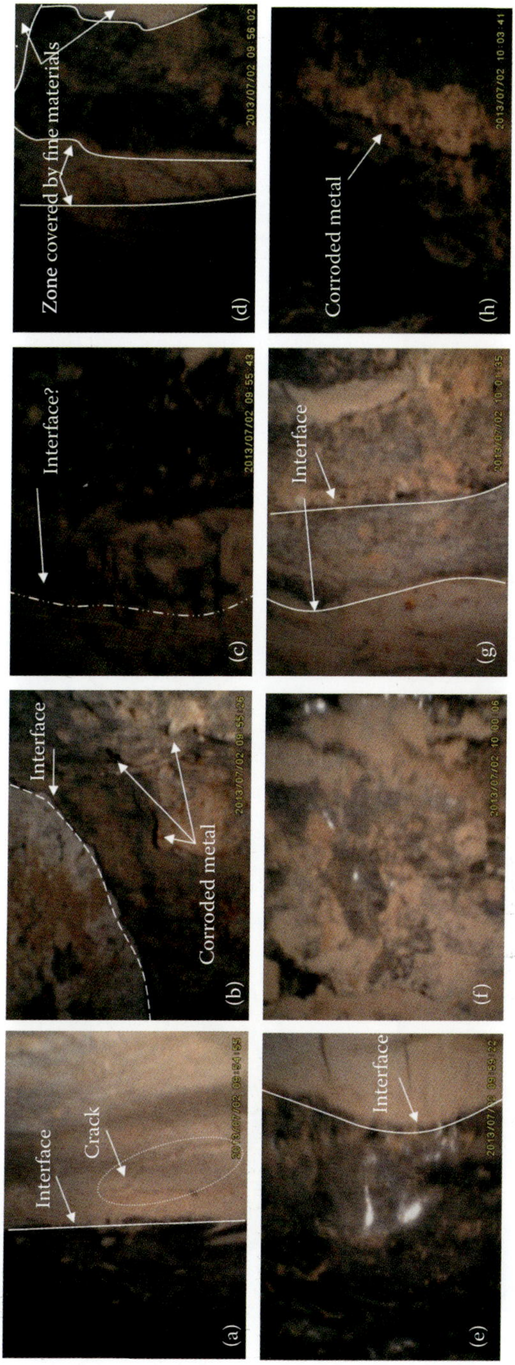

FIGURE 8.9 (a–h) Images resulting from the endoscopy, showing different parts of the column.

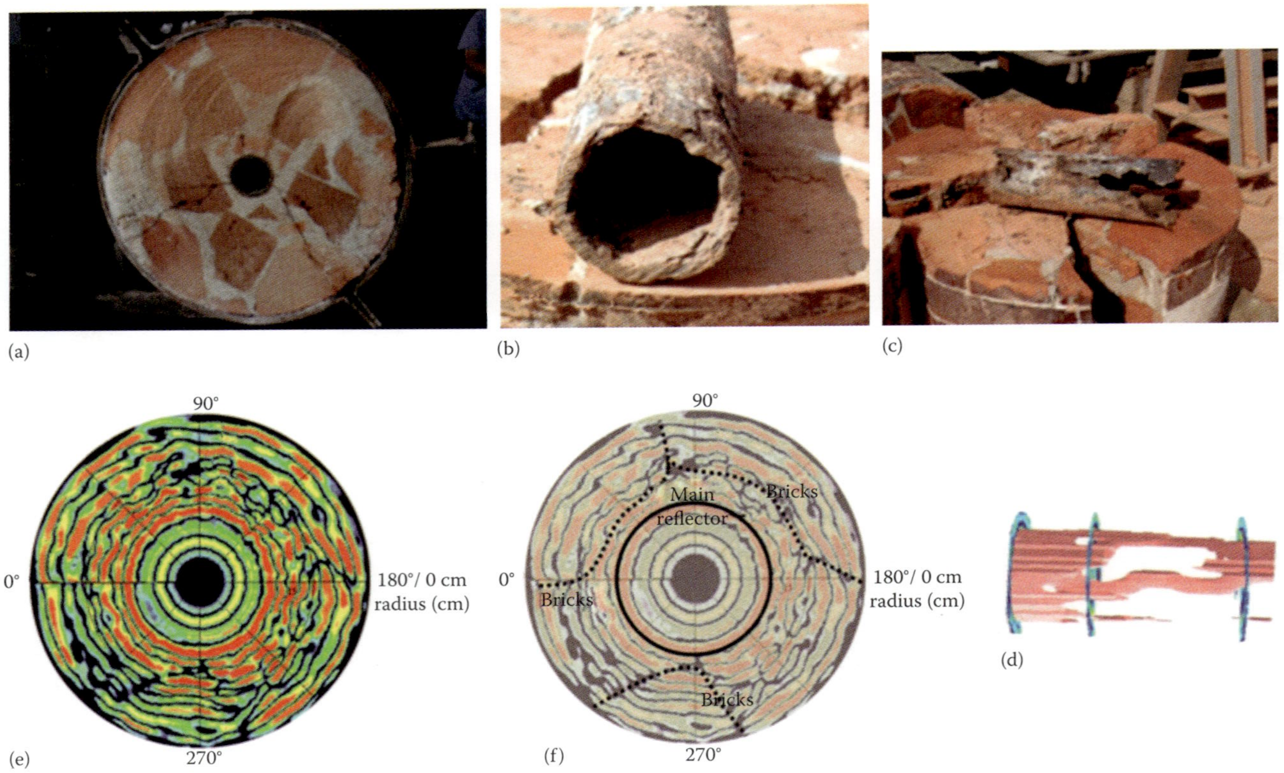

FIGURE 8.10 (a) Cross cut of the column, showing the irregular arrangement of the bricks. (b) Internal drainage tube, cause of damage in the column due to corrosion and humidity. (c) Detail of the column and the corroded part of the tube. (d) GPR 3D image of the tube. (e) Wrapped B-scan showing the heterogeneous anomalies. (f) Possible radar data interpretation.

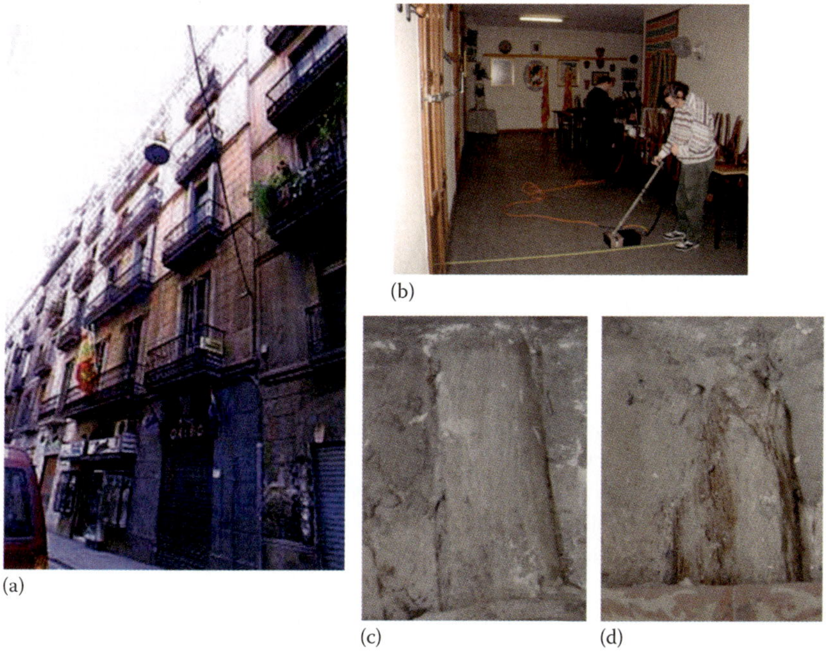

FIGURE 8.11 (a) Assessed building. (b) Radar data acquisition inside the building. (c) Head of beam in correct state of conservation. (d) Damaged head of beam.

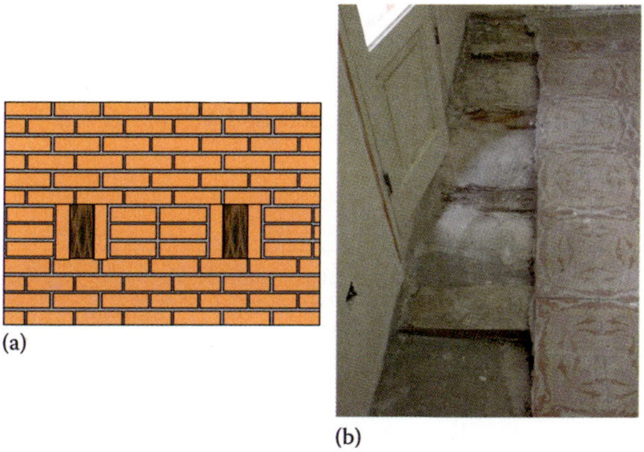

FIGURE 8.12 Head of beams connected to the load-bearing wall. (a) Scheme showing the brick boxes at the connection. (b) Photograph of the head beams connected to the wall.

were uncovered. The second was a radar data acquisition in two radar lines parallel to the wall. Where the radar lines crossed, the beams allowed their exact position to be determined. The third stage was the radar data acquisition along lines placed over the beams. The fourth stage was a destructive test in the zones selected by means of GPR results.

GPR data was processed with a time-zero correction, a dewow, a bandpass filter and a three-point gain function.

8.3.3 Results

GPR data obtained at the radar lines parallel to the wall clearly demonstrate the position of each of the beams (Figure 8.13). As occurs in many cases, the beam section is rectangular, and the hyperbolas caused at the edges of the beams are clearly shown in the radargram. The flat upper part of the beam is detected on the radargrams as a flat anomaly, bordered with the hyperbola branches. Those branches were used to determine an average wave propagation velocity of about 9 cm/ns.

The images also enabled the distance to the ceiling of the upper level to be determined, at about 6 ns. This anomaly was later used to associate one of the anomalies with the images obtained over the beams to the ceiling under each beam.

The radargrams obtained during the third stage of the study were in radar lines along the beams. In these images, irregular anomalies in the zone corresponding to a beam were associated with possible damage, while homogeneous images were associated with beams in a good state of conservation. The extension of the anomalies was used to define possible degrees of damage in each beam. High extension of the anomalies is considered to be an indicator of the beams most likely to be highly damaged. These beams needed further analysis, even being evaluated directly with destructive tests. Figure 8.14 shows the images obtained in three different beams. Figure 8.14a corresponds to a beam in a good state of conservation. Reflections on the beam and the ceiling under the beam are clearly visible, in both cases being regular anomalies recorded along all the radar lines. At the beginning of the B-scans, the inclined anomalies correspond to clutter caused by the wall, the anomaly corresponding to the beam being observed in all this area, which most likely indicates that the support of the beam in the wall is adequate and well preserved. Figure 8.14b also

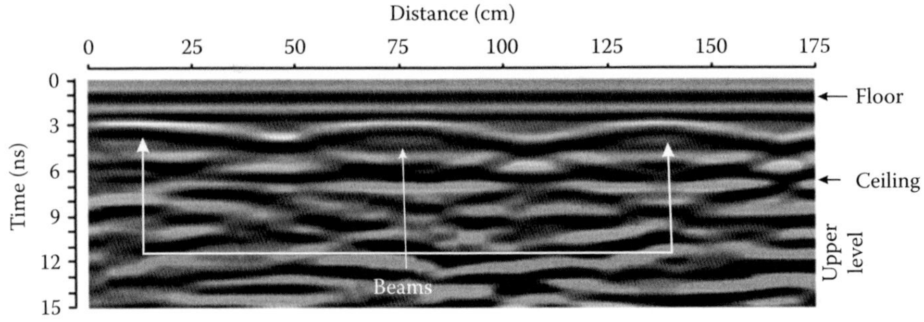

FIGURE 8.13 Radar data obtained in a radar line that crosses the beams.

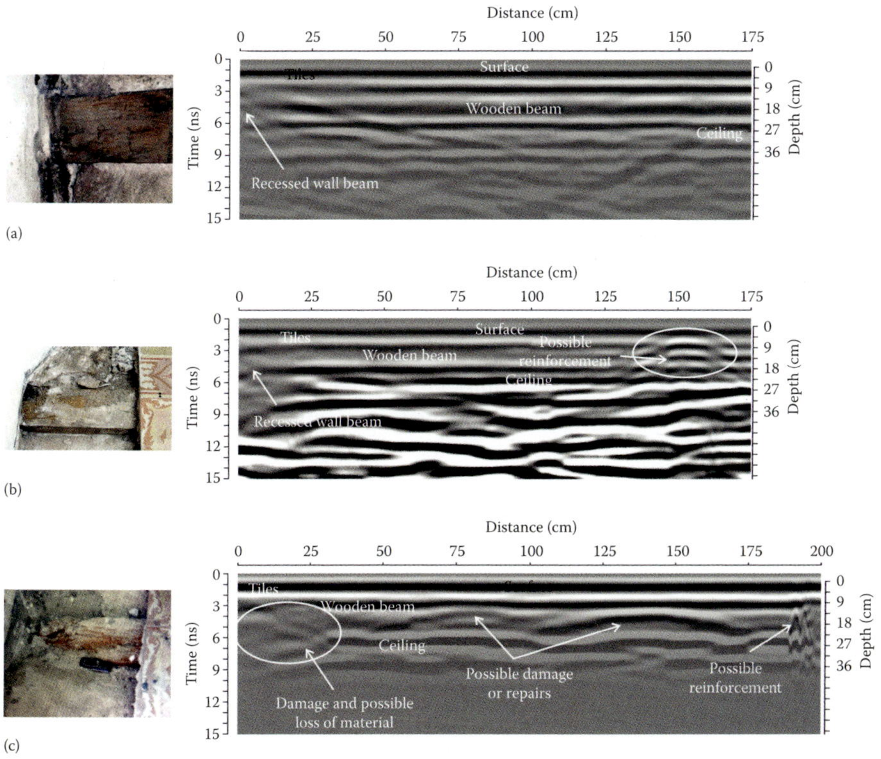

FIGURE 8.14 B-scans obtained along the beams. The photographs correspond to the junctions of each beam with the wall after lifting the tiles in a later study. (a) Radar data obtained in a well-preserved beam. (b) Radar data from a well-preserved beam, probably with a repaired area. (c) Beam poorly maintained, probably with problems of material loss in the junction with the wall.

corresponds to a beam with an adequate and well-preserved wall support but with an old repair to the beam. Finally, Figure 8.14c corresponds to a damaged beam. At the beginning of the B-scan, the reflection corresponding to the beam disappeared in a large area, most likely indicating that the support is not properly preserved and that the beam has probably suffered material loss. This anomaly highlights a zone that could be significantly damaged. On the other hand, the reflection on the beam is not constant, most likely indicating possible damage or even loss of material in different zones.

The analysis of all the radar images provided a map of each room where the most problematic areas could be marked for further analysis or to proceed to the repair of the damage. It is noticeable that differences due to the existence of damage caused by woodworm is not detected using the common offset method. Only in the case of high-extension material loss can the damage and anomalies be clearly observed in the B-scans. Therefore, this study indicates those zones that need immediate repair, but it is necessary to combine the study with a more intensive evaluation using other methodologies.

8.4 CONCLUSIONS

GPR is an NDT methodology that could be used to support studies of structures and cultural heritage buildings. It is useful to detect anomalous zones inside structural members, also allowing the shape and geometry of some of those anomalous zones to be determined after a 3D analysis, requiring careful radar data processing. However, it is necessary to know that these anomalous zones only refer to contrasts in electromagnetic properties; it is difficult, or even impossible in many cases, to identify the materials of the cause of the contrasts between the electromagnetic parameters. On the other hand, accurate analyses usually require a precise determination of the wave velocity. This is possible using previous tests with other methods to determine the depth of any anomaly. Other methods to determine velocities are based on comparing amplitudes – although this is difficult in the case of complex media or a large number of layers, and changes in amplitude can be caused by many effects – and evaluating the hyperbolas caused by refraction of the energy. However, in this last case, the velocity can be overestimated because of the size and shape of the target that caused the refraction. These are problems that are present in all indirect analysis, forcing a combination of methodologies to reduce uncertainties and to obtain a more accurate interpretation.

REFERENCES

Binda, L., G. Lenzi, and A. Saisi. 1998. NDE of masonry structures: Use of radar tests for the characterisation of stone masonries. *NDT& E International* 31 (6): 411–419.

Binda, L., L. Zanzi, M. Lualdi, and P. Condoleo. 2005. The use of georadar to assess damage to a masonry bell tower in Cremona, Italy. *NDT& E International* 38 (3): 171–179.

Diamanti, N., A. Giannopoulos, and M. C. Forde. 2008. Numerical modelling and experimental verification of GPR to investigate ring separation in brick masonry arch bridges. *NDT & E International* 41 (5): 354–363.

Domènech i Montaner, L. 1916. *Hospitales de la Santa Cruz y de San Pablo: Memoria, planta y perspectiva general del proyecto*. Hospital de Sant Pau. (In Spanish)

González-Drigo, R., V. Pérez-Gracia, D. Di Capua, and L. G. Pujades. 2008. GPR survey applied to modernista buildings in Barcelona: The cultural heritage of the College of Industrial Engineering. *Journal of Cultural Heritage* 9 (2): 196–202.

Goodman, D. 2004. *GPR-SLICE. Ground Penetrating Radar Imaging Software. User's Manual.* Geophysical Archaeometry Laboratory, California.

Goodman D., S. Piro, Y. Nishimura, K. Schneider, H. Hongo, N. Higashi, J. Steinberg, and B. Damiata. 2008. GPR archaeometry. In *Ground Penetrating Radar Theory and Applications*, ed. H. M. Jol, 479–508. Elsevier.

Leucci, G., R. Persico, and F. Soldovieri. 2007. Detection of fractures from GPR data: The case history of the cathedral of Otranto. *Journal of Geophysics and Engineering* 4 (4): 452–461.

McCann, D. M. and M. C. Forde. 2001. Review of NDT methods in the assessment of concrete and masonry structures. *NDT& E International* 34 (2): 71–84.

Martinho, E. and A. Dionísio. 2014. Main geophysical techniques used for non-destructive evaluation in cultural built heritage: A review. *Journal of Geophysics and Engineering* 11 (5): Article number 053001.

Michelena, R. J., F. Muir, and J. M. Harris. 1993. Anisotropic travel time tomography. *Geophysical Prospecting* 41: 381–412.

Nuzzo, L., G. Leucci, and S. Negri. 2009. GPR, ERT and magnetic investigations inside the Martyrium of St Philip, Hierapolis, Turkey. *Archaeological Prospection*, 16(3), 177–192.

Pérez-Gracia, V., J. O. Caselles, J. Clapés, G. Martinez, and R. Osorio. 2013. Non-destructive analysis in cultural heritage buildings: Evaluating the Mallorca cathedral supporting structures. *NDT & E International* 59: 40–47.

Pérez-Gracia, V., D. Di Capua, O. Caselles, F. Rial, H. Lorenzo, R. Gonzalez-Drigo, and J. Armesto. 2011. Characterization of a romanesque bridge in Galicia (Spain). *International Journal of Architectural Heritage*, 5(3), 251–263.

Pérez-Gracia, V., R. González-Drigo, and R. Sala. 2012. Ground-penetrating radar resolution in cultural heritage applications. *Near Surface Geophysics* 10 (1): 77–87.

Pérez-Gracia, V., S. Santos-Assunçao, O. Caselles, J. Clapés, and J. A. Canas. 2014. Study of wood beams in buildings with ground penetrating radar. In *Ground Penetrating Radar (GPR), 2014 15th International Conference on* (pp. 31–35). IEEE.

Sandmeier, K. J. 2003. *REFLEX. Program for Prospecting and Interpretation of Reflection and Transmission Data*. Sandmeiered.

Santos-Assunçao, S. 2014. Ground penetrating radar applications in seismic zonation: Assessment and evaluation. Ph.D. diss., Polytechnic University of Catalonia, Barcelona-Tech, Barcelona, Spain.

Santos-Assunçao, S., V. Pérez-Gracia, O. Caselles, J. Clapes, and V. Salinas. 2014. Assessment of complex masonry structures with GPR compared to other non-destructive testing studies. *Remote Sensing* 6 (9): 8220–8237.

Solla, M., H. Lorenzo, F. I. Rial, and A. Novo. 2011. GPR evaluation of the Roman masonry arch bridge of Lugo (Spain). *NDT & E International* 44 (1): 8–12.

9 Survey and Characterisation of Founder's Chapel Pathologies of the Santa Maria da Vitoria Monastery, Batalha

Flávio Vazão, Florindo Gaspar and Hugo Rodrigues

CONTENTS

9.1 INTRODUCTION

The Monastery of Batalha is a remarkable symbol of Portugal, not only because it is a historical landmark, but also because of its great engineering and architectural construction.

The quality of a monument varies depending on the construction techniques, the architecture and the materials used, as well as the interventions and conservation actions taken over the years. All these factors are decisive for the durability and state of conservation.

Stone, which is one of the essential construction materials of the Monastery of Batalha, is subjected to changes over time. For its conservation and restoration, it is necessary to know its characteristics and properties as well as techniques and products for rehabilitation.

There is a growing awareness of the preservation and conservation of ancient buildings, mainly monuments, leading to the development of studies on conservation and rehabilitation of stone materials.

The objective of this work was the study and inspection of the Founder's Chapel, of the Monastery of Batalha, evaluating its state of conservation. The choice of the Founder's Chapel was due to the partially disintegrated construction of the remaining building (having only one common wall with the church), because its construction had not been planned in the initial project.

The development of the study was based on visual evaluation by observing the existing pathologies, performing their mapping and trying to frame and interconnect the possible causes/effects. To support this work, a campaign of nondestructive tests was carried out; namely, the measurement of ultrasonic propagation, thermography and humidity, with the equipment available at the Polytechnic Institute of Leiria. Laser scanning of the Founder's Chapel interior was also carried out in collaboration with TOPCON Corporation.

9.2 THE MONASTERY AND THE FOUNDER'S CHAPEL

The Monastery of Batalha was ordered to be built by King João I in 1388. It came about as a result of a promise made to the Virgin Mary that, if he won the decisive battle against the Castilian troops, he would ascend to the throne of Portugal (hence the name attributed to the Monastery of Santa Maria da Vitória), thus guaranteeing the independence of Portugal. This was the Battle of Aljubarrota, which took place on 14 August 1385, and the construction of the monastery began one or two years later.

The symbol of the Portuguese and European history of art that is the Monastery of Batalha was labelled as one of the greatest religious complexes of Portuguese Gothic style, thus influencing the buildings in Portugal at that time (Dias 1994). The monastery has been a national monument since the nineteenth century. It is currently under the responsibility of the Direção Geral do Património Cultural (DGPC) and has been part of the UNESCO World Heritage List since December 1983 (Andrade 1989).

The Founder's Chapel was designed by the architect Huguet and built in the second quarter of the fifteenth century. It is a chapel designed to serve as the royal pantheon and contains several tombs of the most important figures in the history of Portugal. Among them we can find King João I and his wife Filipa de Lencastre's joint tomb in the centre, and those of their children (Infante Henrique, Fernando, João and his wife Isabel, and Pedro and his wife Isabel de Urgel) lining the south wall. Lining the west wall are the tombs of King Afonso V, João II and his son Afonso.

Located in the southwest part of the church, we can find the Founder's Chapel, in which we can see the tombs described previously and also eight columns supporting the roof. It has a square plan, and after the first peripheral roof level of four hips, an octagonal roof with eight hips rises over the columns, establishing a second roof level enclosed by a ribbed dome with a star shape (Branco and Azevedo 1990).

9.3 MAPPING OF PATHOLOGIES IN THE FOUNDER'S CHAPEL

The elaboration of a graphical representation indicating the zones where there are manifestations of pathologies, damages or degraded elements, with an indication of their intensity, allows us to have a better perception of the state of conservation and also gives us a better understanding of its possible causes. This procedure involved the three outer elevations and the roof of the Founder's Chapel. Due to the extension of the mapped area and the problematic close observation of the high-level zones, it was difficult to achieve rigorous detail.

The mapping was divided into several types of degradation, as follows: sanding, biogenic crusts, patina, moisture, opened joints, broken stones, biological colonisation, dry moss, soiling, alveolisation, peeling and cracks.

Recent or recently cleaned stones, which had also been included, although not as a pathology, had remarkably different characteristics, such as greater susceptibility to degradation. For sanding, black crusts and patina degradation types, they were divided into two grades: Grade 1 for higher intensity and Grade 2 for lower intensity.

The mapping was done in a chromatic form. The colours assigned to the sanding process were dark red for Grade 1 and light red for Grade 2, and for biogenic crusts, it was dark grey for Grade 1 and light grey for Grade 2. The colours chosen for the patina were dark orange for Grade 1 and light orange for Grade 2. Dark blue was chosen for moisture, and light blue represented the opened joints. The dark brown colour was chosen for broken pieces, green for biological colonisation, pink for recent or recently cleaned stones, and blue (ocean tone) for soiling. Violet represents alveolisation, blue (sky tone) represents peeling, and lastly, purple represents cracks.

Figure 9.1 shows the pathologies of the east, south and west elevations and the roof plan of the Founder's Chapel.

The most common pathology in stone is biogenic crusts. These are black layers that cover the monastery due to the proliferation of biological colonisation, such as lichens, fungi and microorganisms. The biogenic crusts are present on the pinnacles and in the upper part of the flying buttresses, fleurs-de-lis and buttresses mainly below the gargoyles. The location of this pathology is mainly due to areas in which the water runoff caused microorganisms to be transported by the water and then deposited in the stone.

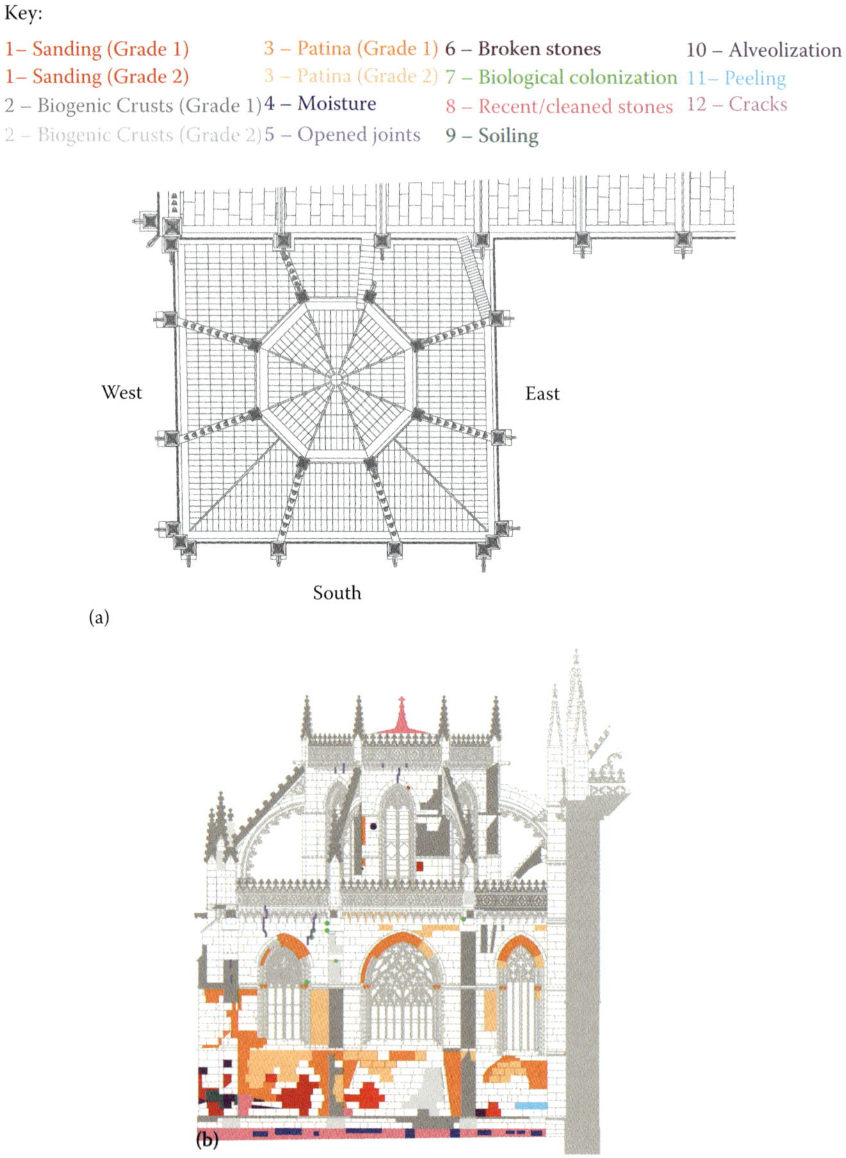

FIGURE 9.1 (a) Layout of the elevations of the Founder's Chapel; (b) mapping of east elevation pathologies.

In general, the patina observed in the three elevations is in more protected areas. The consolidation of the stone with waxes and oils in old conservation campaigns may have originated this pathology, which, in areas less exposed to the environment, could have been conserved in the stone, unlike those that had been more exposed and in time dissolved. This phenomenon is more visible in protected ogive windows and walls near and between the buttresses.

FIGURE 9.1 (Continued) (c) mapping of south elevation pathologies; (d) mapping of west elevation pathologies.

In the southern elevation, the humidity in the first row of the wall is remarkable. Although it has been repaired in the three elevations of the chapel, it shows signs of absorption and infiltration of water. It is not known whether this finishing layer was applied at the time of the construction works done to lower the outside level of the pavement or afterwards. In the first stone layer of the east side, as in others, it is clearly visible that it is only a 3–4 cm plate. The high humidity found in the first layer of the south wall may be due to a significant amount of ground water, as there had been a rivulet near the chapel in the past. The stains on the stone may also be due to the absorption of salts, chromatically altering the stone. Having remained buried for a few centuries, the masonry collected the saline material from the earth.

The absorption of water by the roof before the last works led to the chromatic alteration of the ceiling in the chapel to an orange colour. This phenomenon is observed with more relevance in the central zone. The colour is certainly due to the infiltration of the fines from ceramic bricks, which were used to fill the support of the roof tiles and surround the tension bars.

The generality of the sanding and peeling occurs up to about 3 metres in height. It is suspected that most of them are not original stones from the initial construction of the monastery, probably being from nineteenth-century restoration works. In some places, there are other stones that have apparently been changed. It is supposed that they are only finishing plates that have been set in the same way as in the first row.

The flying buttresses, besides their structural function, also have the function of draining the roof waters to the gargoyles. Old obstructions in the drainage system have led to water infiltration in the roof, which has accelerated the degradation of the flying buttresses, causing sanding in the intrados and biological colonisation, vegetation and biogenic crusts in the extrados. These pathologies could have been accelerated in the flying buttresses due to the inferior quality of the stone used in the restoration work.

The opened joints are visible mainly in the central part, above the windows in the octagon and also in some windows of the lower level. As will be presented in the laser scanning analysis, this may be due to the displacements at the top of the octagon. To control the thickness of these openings, a monitoring system needs to be put in place. There is, however, the surrounding metallic strapping around the Founder's Chapel, which helps to confine the roof. In it are vertical spikes and crossed and perimeter tension bars. Probably, the tension bars cross the octagon from one end to the other, passing through the roof. The tension bars are attached to the spikes through a ring-shaped connection.

9.4 NONDESTRUCTIVE TESTING CAMPAIGN

In a historic monument such as the Monastery of Batalha, the diagnosis should preferably be supported with nondestructive tests. The tests carried out were the evaluation of the propagation of ultrasonic waves in the stone, the analysis of the surface temperatures through thermography, and the measurement of the surface hardness of the stone with the Schmidt hammer. In the stones analysed with ultrasound, the humidity levels were also measured.

9.4.1 Ultrasound Testing Campaign

9.4.1.1 Materials and Methods

In the ultrasound test, a coupling agent is used to ease the transition of the ultrasound between the transducers and the material under analysis. In the present case, the coupling agent was selected with special care so as not to damage or stain the stone found in the monastery. After preliminary laboratory tests, a modelling clay material was chosen, since it left no sign of its usage and was easily removed, unlike petroleum jelly, which penetrates into the porosity of the stone. During the tests, care was taken to renew the modelling clay material used as coupling agent when it began to dry after a set of measurements to obtain stability in the results.

In each stone tested, five readings were taken along a certain straight path, using the indirect method, with successive readings at intervals of 15, 25, 35, 45 and 55 cm. To get better precision and speed in the measurements, a specifically made gauge was used, which also helped in the placement of the transducers. One of the transducers was fixed, and the other transducer was moved along according to the various readings.

The tests were carried out looking for stones with different pathologies such as biogenic crusts, patina or moisture, avoiding the stones with surface damage, because it is difficult to collect data on these stones. However, some dispersion is expected in the results, as in addition to the nonperfect coupling, there are still other variables, such as the orientation of the façade, the various pathologies and the location. Nevertheless, tests were undertaken to homogeneously test the pathologies for the different orientations.

To determine the relationship between the results obtained with petroleum jelly and with the modelling clay material, tests were carried out on five stones. The tests were performed with both coupling agents at the same points. The stones analysed were a block of marble with a thickness of 58 cm, a block of limestone with a thickness of 38 cm and another three stones of limestone, of which two had a thickness of 5 cm and the other had a thickness of 3 cm.

The tests were done on the outside walls of the chapel, consisting of the west, south and east elevations, on the inside, and on the octagonal envelope above the first level of the roof.

Measurements were taken at 10 points in the east, 14 points in the south and 13 in the west elevation. On the inside, 46 measurements were made, distributed over the walls, columns and tombs: 1 measurement in the corner of the south wall with the east wall, 10 on the east wall, 19 on the north wall and 3 on the west wall, 5 on the tombs of the west side, and the remaining 8 on the central columns of the chapel.

In the octagonal envelope, 22 measurements were made. There were also 17 measurements on the wall on the south side of the church to compare with the measurements made on the chapel.

The most recent or recently cleaned stones are also of interest to the analysis, as they were removed from quarries different from those used in the initial construction of the monastery and are suspected to only be 3–4 cm in thickness.

The tests were carried out up to 3 metres above the height of the ground. The measurements were evenly distributed on the outer perimeter of the Founder's Chapel in both the lower and the upper levels.

9.4.1.2 Influence of Transducer Distance on Speed

The overall mean value of the ultrasound velocity obtained using the modelling clay was 2362 m/s. The individual mean velocities for each distance of the transducers are 2395, 2404, 2348, 2341 and 2337 m/s for 15, 25, 35, 45 and 55 cm of spacing, respectively. From this analysis, it is verified that the differences are not very different from each other. However, with the exception of the first value, the speed decreases slightly as the distance increases. This loss of speed may be due to the increase in dispersion of the ultrasonic waves, causing transmission loss when passing though the stone.

9.4.1.3 Coupling Agent Influence

Figure 9.2a presents the results of the comparison between the petroleum jelly and the modelling clay.

The results obtained allow us to conclude that the use of the modelling clay results in the ultrasound propagation time being higher than for the petroleum jelly, as would be expected, since it does not provide such effective transmission. In addition, it was verified that the dispersion of values is lower for the results obtained with petroleum jelly.

From the analysis of the ultrasound results, no difference seems to exist due to the different stone thicknesses.

The values of the propagation time increase as the distance between the transducers increases as a result of the increased dispersion of the ultrasound. For each distance (15, 25, 35, 45 and 55 cm), the quotients between the average values of the transmission times obtained with petroleum jelly and modelling clay are 0.410, 0.495, 0.519, 0.556 and 0.591, respectively. This means that the transmission times obtained with the petroleum jelly are about twice the times for modelling clay.

Using the previous coefficients, the calculation give an expected average velocity for petroleum jelly of 4657 m/s, varying between 3578 and 5713 m/s in 90% of the cases.

Compared with other work, 4657 m/s is lower than the 6300 m/s obtained by Kahraman et al. (2007) on limestone blocks recently extracted from the quarry. That could be a reason for the higher velocity obtained. The density and compressive strength of limestone analysed by Kahraman et al. (2007) were 2660 kg/m^3 and 100.9 MPa, respectively, values higher than the stone of Batalha Monastery, which has values of 2340 kg/m^3 and 37 MPa, respectively (Branco and Azevedo 1990).

Bromblet et al. (2012) present average values similar to those of the Founder's Chapel, with 4715 m/s for the north zone of the gallery and 4500 m/s for the south zone of the gallery, using the air chambers of bicycles as the coupling agent in the Church of Saint-Trophime in Arles, France. It can be concluded that the ultrasound values obtained are in line with those obtained in other works.

9.4.1.4 Estimating the Modulus of Elasticity

The dynamic modulus of elasticity was estimated according to the following equation relating the dynamic modulus of elasticity with the Poisson coefficient, density and ultrasonic velocity (estimated for petroleum jelly coupling agent):

$$E = (1 + \upsilon)(1 - 2\upsilon) / (1 - \upsilon)\rho * V^2 \qquad (9.1)$$

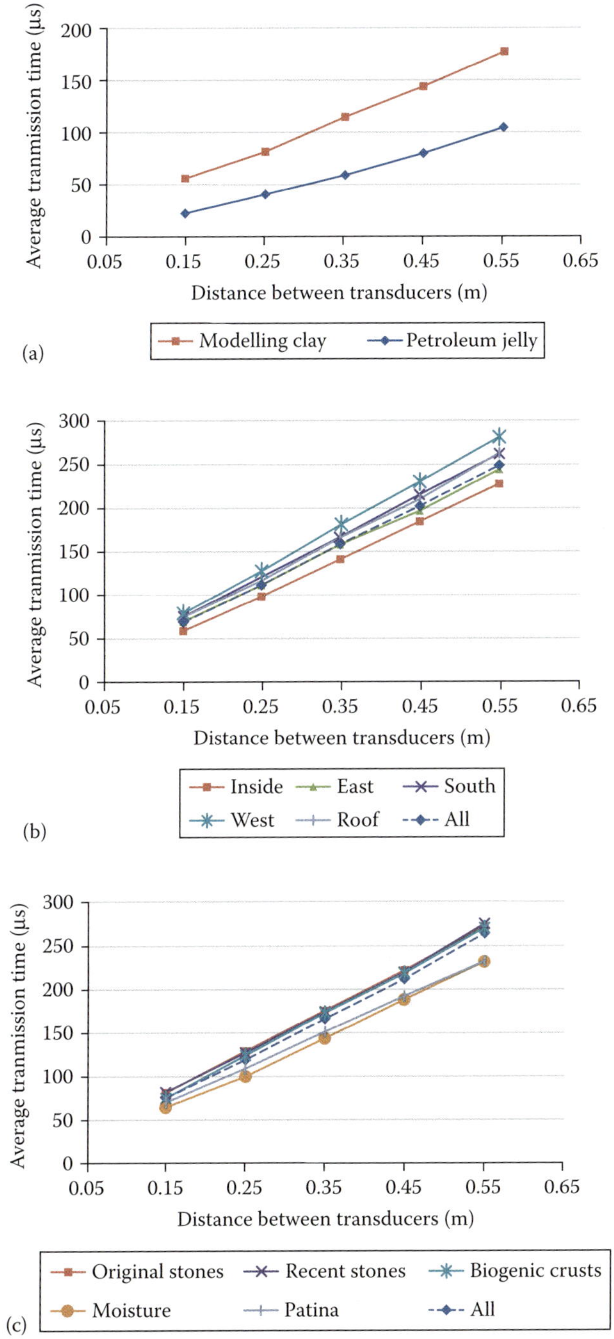

FIGURE 9.2 (a) Comparison between the average transmission times obtained with petroleum jelly and with modelling clay; (b) comparative diagram between orientations/location; (c) comparative diagram between pathologies.

where:

$E =$ dynamic modulus of elasticity (kN/m^2)
$\rho =$ density (ton/m^3)
$\upsilon =$ Poisson coefficient
$V =$ ultrasonic velocity (m/s)

The density considered was based on the value indicated in the inspection work and the structural evaluation at the Monastery of Batalha done by Branco and Azevedo (1990), whose value is 22.94 kN/m^3. According to Guimarães (2009), the Poisson coefficient for the stone is estimated to be between 0.20 and 0.30, so the average value of 0.25 was taken into account. Thus, the estimated value for the dynamic modulus of elasticity was 41.6 GPa. The value converted to the static modulus of elasticity, according to Najibi et al. (2015), was 20.9 GPa, according to the following equation:

$$Es = 0.14 \times Ed^{1.96} \tag{9.2}$$

where:

$Es =$ static modulus of elasticity (kN/m^2)
$Ed =$ dynamic modulus of elasticity (kN/m^2)

In this estimation of the modulus of elasticity, some sources of error are associated, such as the accuracy of the Poisson coefficient, the conversion of the values obtained with the modelling clay to the petroleum jelly-related values, and errors in the measurement of the ultrasonic velocity with the modelling clay, given the difficulty of getting perfect ultrasound transference between the transducers and the stone. In addition, there is the possibility that the stone has lost its properties over time, mainly on the outer surfaces, leading to higher resistance to ultrasound propagation.

9.4.1.5 Influence of Orientation and Location

In the analysis of the influence of the stones' orientation and location, as well as the influence of the pathologies, the time and velocity values were used for the modelling clay as a coupling agent.

The average transmission time values obtained by orientation and location are shown in Figure 9.2. The figure includes the average times inside the chapel, outside (east, south and west elevations) and in the roof (the envelope of the octagon). The dashed line is the global mean.

The conclusions drawn from Figure 9.2 are as follows:

- The averages from the roof and the south elevation have the same magnitude of ultrasound propagation time.
- The ultrasound propagation times of the east and south elevations and the roof are equivalent to the global average.

- Regarding the orientation, we can conclude that the stone from the west has the shortest transmission time. This may be due to the higher exposure to atmospheric agents and also to the higher exposure to pollution caused by the constant flow of traffic on the nearby main road, which might have led to further degradation of the stone.

9.4.1.6 Influence of Pathologies

The pathologies considered were patina, moisture and biogenic crusts. In addition to these, the original stones with no pathologies and the recent stones have been incorporated. Although the recent stones have no pathologies, they have different properties from other stones. As in the Figure 9.2, the total average is also presented (Figure 9.2). The following conclusions were obtained:

- The mean times for biogenic crusts, stones without pathologies and recent stones have similar values and are generally above the average.
- The shortest times are relative to patina and humidity. The water in the stone, as already mentioned, helps in the propagation of ultrasound, because it fills the empty air pockets. The patina contributes to the fast propagation, probably because it creates an adherent layer to the stone, reducing the outer porosity and protecting the stone material.
- All variables show proportionality when referring to the distance/time relationship.

9.4.2 Thermography

9.4.2.1 Testing Campaign

Thermography is a technology that combines visual and infrared images. The equipment used in thermography was a Fluke Ti45FT.

The reading and editing of images were performed using SmartView 3.1 software. In the analysis of the results, the value of 0.95 was used for the emissivity according to Mendonça (2005). The objective was to detect the pathologies. However, some stone anomalies, such as small cracks, are only possible to find with drying effects (Silva 2012) using dynamic drying and cooling.

The tests were carried out homogeneously all over the Founder's Chapel, covering all elevations, the roof and the interior and analysing the influence of other variables such as the age of the stone and pathologies, trying to identify differences in temperature due to the various variables.

The test campaign took place in the morning, when the stone was still warming up.

9.4.2.2 Results and Discussion

From the analysis of the images collected with the thermographic camera, several conclusions were obtained. It is remarkable to see the low temperature of the stone in the dark areas, because in the morning the sun quickly warmed the exposed areas.

It was observed that the temperature was higher in the stones with the darker surface due to their higher solar radiation absorption in contrast to the lighter stones, which reflected it. These dark surfaces in the stone are due to the biogenic crusts or patina, as shown in Figure 9.3.

In some places, differences in the surface temperature of some stones were observed, as shown in Figure 9.3. It is believed that these stones were replaced and have different behaviour in comparison with the surrounding stones, as is emphasised by the thermogram showing the supposed newer rocks with cooler temperatures (represented with a blue colour). These stones probably have a lower thickness and do not have the width of the entire wall. The cooler temperature of the stone windowsill is due to its location and inclination, which limit the exposure to solar radiation, so this is probably an original stone.

A high temperature was found in the walls near the floor, which can be explained by the presence of humidity, which accelerates the heating and leads to a higher temperature, as shown in Figure 9.3c.

Cooler surfaces were also found in stones that suffer from peeling, sanding and alveolisation, which may be due to the lower compacity and the irregular surface, which resists the penetration of solar radiation.

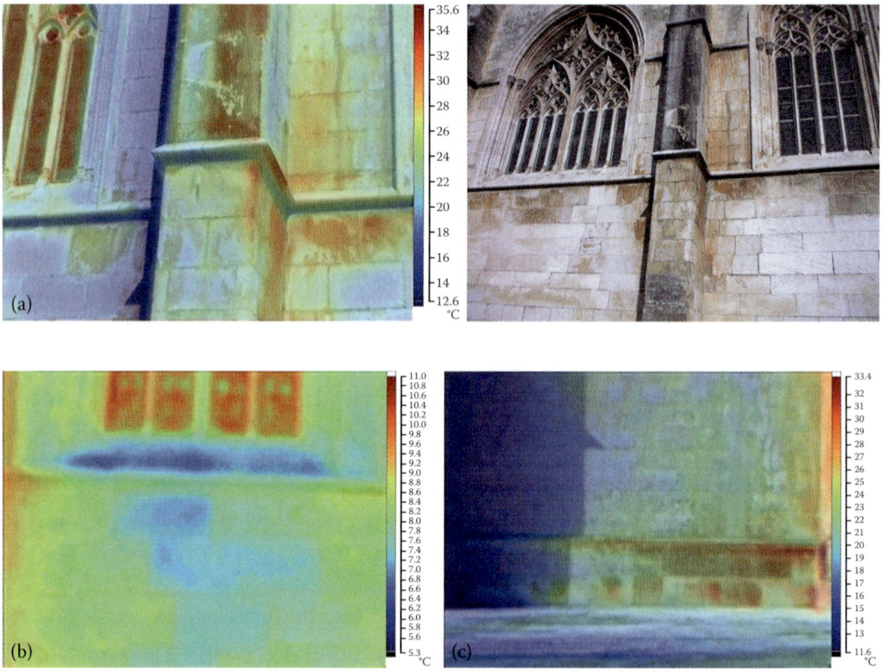

FIGURE 9.3 (a) Biogenic crusts thermogram and corresponding location photo; (b) thermogram of recent stones; (c) heating thermogram next to the floor.

9.4.3 Humidity Measurements

For each stone analysed with ultrasound, the humidity was measured. The range of moisture values was from 0.1% to 6.9%, with an average of 4.4%. Comparing the inside and outside humidity, the average was equal to 5.0 and 4.1, respectively. The higher humidity in the interior is due to absorption and condensation, with the drying and evaporation in the interior being slower than outside. The sun and wind outside provide faster drying.

9.4.4 Determination of Rebound Numbers

9.4.4.1 Testing Campaign

For the testing campaign carried out with a Schmidt hammer, several stones were analysed. They were chosen from among those previously analysed with ultrasound, homogeneously scattered throughout the chapel. As with the ultrasound tests, measurements were not taken on stones with surface damages, because these pathologies would influence the hardness, giving unreliable values of rebound numbers. For each stone, five measurements were collected.

9.4.4.2 Results and Discussion

The values of each set of five measurements range between 30.7 and 59.2, which are not very distant from the values collected by Machado (2012) on the analysis of the granite from the Cathedral of Lamego.

The standard deviation of the measurements for each stone varied between 0.87 and 4.64. Although only five measurements were collected on each stone, there were no big variations.

It was verified that the surface hardness is higher inside the Founder's Chapel. On the outside, the rebound numbers are lower, probably due to ageing by atmospheric agents along with pollution.

The wall on the north side, common to the church, has the highest rebound numbers, probably due to the fact that it is an interior wall and is not subject to atmospheric factors, unlike the exterior walls.

It was found that the rebound numbers for the different orientations of the chapel agree with the ultrasound propagation velocity. The highest value of rebound number was obtained in the interior, decreasing in the east elevation, followed by the roof and the south elevation, and finally a lower value for the west elevation (Figure 9.4). This sequence makes sense, since it agrees with the degree of exposure to ageing agents such as pollution and atmospheric agents, the east elevation is the most protected, and the west elevation, located on the main façade, is more exposed to atmospheric agents. In addition, the west elevation faces the busy main road near the monastery, where traffic is an active source of pollution, which might also be one of the causes for the values obtained.

9.4.5 Laser Scanning

Laser scanning was used to verify the verticality of the Founder's Chapel. The SCENE LT software was used, and the coordinates of specific points were taken.

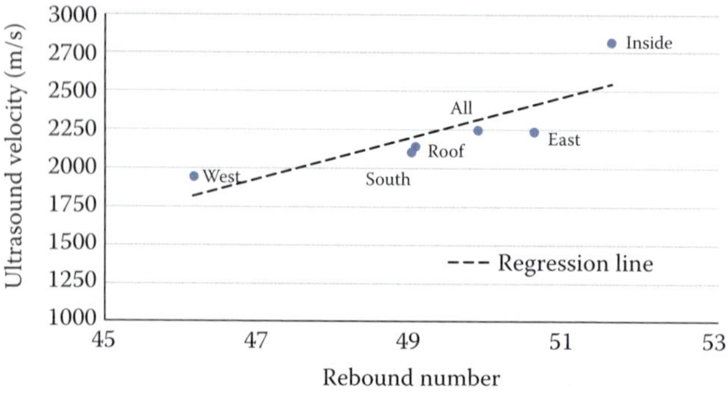

FIGURE 9.4 Comparison between the Schmidt hammer and ultrasound values by orientation.

The interior of the chapel was analysed according to eight vertical alignments corresponding to the eight columns in the central area, which connect the lower with the upper part of the Founder's Chapel, between the windows of the octagon at the level of the roof. In each vertical alignment, the coordinates of four points were taken, at 3.7, 12.7, 13.6 and 16.5 m from the floor. Figure 9.5a shows the shear with an indication of the measured levels.

Figure 9.5b shows the laser scanning measurements at the four levels. The 3.7 m dimension is in the natural scale, and all the others have a factor scale increase of 20 with respect to the coordinate difference to the lower level (3.7 m). It is shown that the Founder's Chapel presents a displacement to the south (maximum of 11.4 cm), east and west (maximum of 8.2 cm) sides, probably due to the consequences of the displacement of the church, or due to single events, such as construction or seismic activity. This phenomenon is more notable as the distance from the floor increases. It is also observed that the part of the octagon on the side of the church was subject

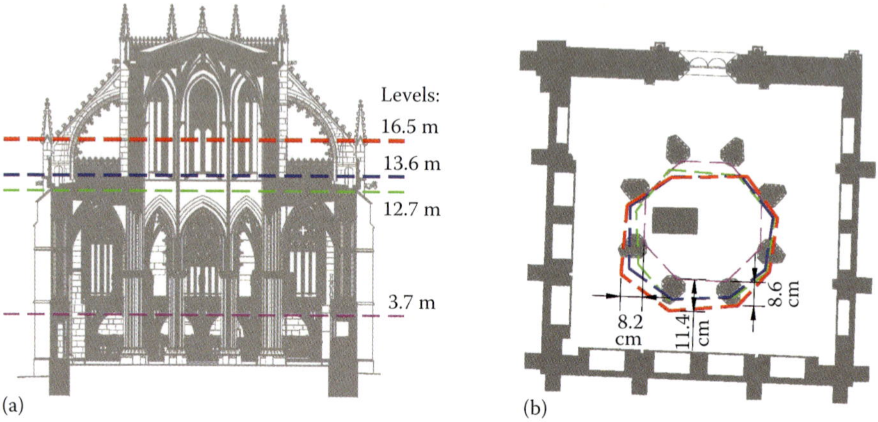

FIGURE 9.5 (a) Shear with indication of the measured levels; (b) laser scanning measurements.

to smaller displacements as a result of the higher rigidity of the structure in that direction.

The results show that the columns have an average vertical inclination of around 8 mm per metre.

9.5 CONCLUSIONS

The emblematic Monastery of Batalha has been suffering serious pathologies in its stones. The stones mostly affected by the degradation seem to be those from the nineteenth-century restoration campaign.

On the exterior of the Founder's Chapel, the biogenic crusts and the patina cover most of the surface of the stone. Sanding and peeling are more problematic pathologies and also exist in this building, but to a lower extent. The biogenic crusts and patina form a layer that covers the outer surface of the stone, helping to provide protection to some extent.

From the measurements taken using laser scanning, it was concluded that the chapel has displacements in the south, east and west directions. The maximum horizontal displacement between the lower and higher levels is about 11.4 cm in the south direction. These displacements are probably the major cause for opened joints above the windows, which apparently was the reason for the implementation of the metallic roof reinforcements. It is possible that the destruction of the roof due to the earthquake of 1755 and the later construction of the new roof have accelerated the opening of the octagon.

The ultrasound and Schmidt hammer test results are in line with each other regarding the characteristics of the stone. On the outside, the west elevation shows the worst results, and the east elevation shows the best results. The interior presents the best results; degradation is minimal, with the exception of some tombs.

In the thermography study, the most relevant outputs were the supposedly more recent stones, which are clearly distinguishable from the others, including the entire first row of the chapel exterior.

ACKNOWLEDGEMENTS

The authors thank the director of the Batalha Monastery, Dr Joaquim Ruivo, as well as all the technicians of that monastery, in particular Dr Pedro Redol, for their availability. In addition, the authors would like to thank Topcon Positioning Portugal, which made the laser scanning possible.

REFERENCES

Andrade, S. 1989. *Mosteiro de Santa Maria da Vitória*. Instituto Português do Património Cultural.

Branco, F. A. and J. Azevedo. 1990. *Mosteiro da Batalha: Inspecção e Avaliação Estrutural*. Report by the Centro de Mecânica e Engenharia Estruturais da Universidade Técnica de Lisboa to the Instituto Português do Património Cultural, Lisbon.

Bromblet, P., V. Vergès-Belmin, and S. Simon. 2012. Ultrasonic velocity measurements for the long-term monitoring of the degradation of marble columns in the cloister of the church of Saint-Trophime in Arles (France). *12th International Congress on the Deterioration and Conservation of Stone*. Columbia University, New York.

Dias, P. 1994. *A Arquitetura Gótica Portuguesa*. Editorial Estampa.

Guimarães, M. I. M. 2009. Caracterização de paredes de Alvenaria de pedra por técnica sónica. Dissertação de mestrado, Faculdade Engenharia da Universidade do Porto.

Kahraman, S., U. Ulker, and M. S. Delibalta. 2007. A quality classification of buildings stones from P-wave velocity and its application to stone cutting with gang saws. *The Journal of the Southern African Institute of Mining and Metallurgy* 107: 427–430.

Machado, B. 2012. Caracterização das litologias e patologias da Sé de Lamego. Um estudo sobre alteração em monumentos graníticos. Tese de Doutoramento em Ciências da Terra e da Vida. Universidade de Trás-os-Montes e Alto Douro, Vila Real.

Mendonça, P. 2005. Habitar sob uma segunda pele. Estratégias para a redução do impacto ambiental de construções solares passivas em climas temperados. Tese de doutoramento, Universidade do Minho.

Najibi, A. R., M. Ghafoori, G. R. Lashkaripour, and M. R. Asef. 2015. Empirical relations between strength and static and dynamic elastic properties of Asmari and Sarvak limestones, two main oil reservoirs in Iran. *Journal of Petroleum Science and Engineering* 126: 78–82.

Silva, D. 2012. Diagnóstico de patologias em fachadas utilizando termografia. Dissertação de Mestrado, Faculdade de Engenharia Universidade do Porto.

10 Dynamic Measurements as Support to the Assessment of the Monastery of Santa Maria Da Vitoria, Batalha

Micael Pedroso, Hugo Rodrigues, Humberto Varum and António Arêde

CONTENTS

10.1 INTRODUCTION

Structural health monitoring (SHM) is a powerful instrument for evaluating the structural behaviour and conservation of the historical patrimony, for which the structural analysis is becoming ever more complex, considering the loads, the ageing of materials and the environmental factors the structures have been exposed to (Antunes et al. 2009, Lima et al. 2008).

One of the main target applications of SHM systems in civil constructions is historical buildings. The identification of defects and damages in existing ancient buildings, as well as their causes, can be very difficult to perform, which makes the rehabilitation intervention a very difficult task. For such types of structures, the collection of data from structural monitoring is often the only way to understand their behaviour and to enable adequate intervention in the structures. One of the possibilities for performing the monitoring over time is based on observation of the changes in the static and dynamic response parameters, combined with measurements of environmental actions (Oliveira et al. 2005).

The dynamic response is an intrinsic characteristic of structures. It is mainly dependent on the material properties, geometry and supporting conditions, and it will remain constant unless otherwise affected by physical changes (due to repair or structural modifications) or damage to the structure (Halling et al. 2001). In fact, the natural frequencies are sensitive indicators of structural integrity. Thus, an analysis of periodical frequency measurements can be used to monitor the structural condition (Salawu 1997, Ju and Mimovich 1988).

Regarding historical structures, dynamic identification tests are carried out to (D10.4-NIKER 2012)

- Obtain information on the global dynamic behaviour of the structure, including natural frequencies, modal shapes and damping coefficients
- Validate and update a structure's numerical model by comparing experimental and numerical natural frequencies and mode shapes
- Assess the level of connection between different parts and identify weak structural features
- Appraise the influence of major damage on the structural response
- Evaluate the effect of possible repair and strengthening solutions by comparing the dynamic parameters before and after the intervention, hence improving the design for future applications in other similar structures
- Allow monitoring before and after the intervention by periodically repeating the dynamic tests and comparing the results
- Study the influence of environmental parameters in the case of buildings sensitive to them, as in particular earthen structures

To obtain, experimentally, the dynamic characteristics of a structure, two distinct methods are available: force vibration testing (FVT) and ambient vibration testing (AVT) (Cantieni 2005).

On the other hand, vibrations imposed on buildings can induce structural and nonstructural damage as well as a sense of insecurity among the buildings' users (Johnson and Hannen 2015). Such vibrations can also accelerate material degradation. In historical buildings, due to the historical value and the difficulty associated with repair, it is clear that if the vibration sources are near these buildings, there is an obvious need to monitor the intensity of the vibrational loads and the behaviour of these structures (Hunaidi 2000). The increase of traffic in the vicinity of historic buildings may impose vibrations for which such structures are probably not prepared. Thus, it is necessary to evaluate the vibrations that affect these structures so as to understand their effect on the monument's behaviour and predict possible damage (Ellis 1987).

Several studies on traffic-induced vibration indicated that the level of measured vibrations did not seem to cause structural damage to building structures (Crispino and D'Apuzzo 2001, Hao et al. 2001, Gupta et al. 2008, Xu and Hong 2008). However, vibration can induce cosmetic and minor damage, and this is undoubtedly of the utmost concern when considering heritage buildings.

The Monastery of Santa Maria da Vitória, best known as the Monastery of Batalha, is one of the symbols of the Dynasty of Avis. It was raised on the initiative of King D. João I, under the direction of the Portuguese architect Afonso Domingues, following a promise made to the Holy Virgin for the victory in the Battle of Aljubarrota

(1385). The monastery has the most important and oldest Portuguese stained glass work, being the earliest example of this art in Portugal. It is considered a treasure, both for the rarity of this art form in Portugal and for its artistic importance and value (Vilarigues et al. 2011, Fernandes et al. 2008, Raguin and Higgins 2008, Redol 2003).

This chapter presents a field campaign of daily vibration measurements of the traffic vibration induced at the monastery. The Monastery of Batalha is situated close to the IC2 road, which has high traffic intensity. Therefore, these vibration measurements provide a good indication of the vibrations affecting the site during the lifetime of the monument.

10.2 IDENTIFICATION OF THE CRITICAL POINTS

To understand the vibration behaviour from its origin (Point A) to the Monastery of Santa Maria da Vitoria, an initial study was developed to understand vibrational behaviour at several points throughout the monastery, to determine whether the front base point experiences the highest vibration levels or whether there is any amplification of the vibrations throughout the monument.

To accomplish this objective, two seismographs were used: one fixed in the ground façade of the monastery (Point O) and another, mobile seismograph placed in various locations. The seismographs were then programmed to perform measurements at the same time, thus allowing their direct comparison for the same environmental actions (Figure 10.1). With this set-up, it is possible to investigate the potential amplification of vibrations along the height of the front façade (Points O, B and C), and an additional point to investigate possible amplification at a point farther away but in a slender element: the tower at Point D (see Figure 10.2).

Measurements were carried out using two seismographs (model GSR-16) with the objective of measuring the accelerations during the study interval. Data acquisition

FIGURE 10.1 Points where the measurements were performed.

FIGURE 10.2 Seismographs placed at (a) Point A; (b) Point O; (c) Point D.

was performed for each point during a 15 min reading, acquired at 250 samples per second. The longitudinal, transverse and vertical axes were defined in accordance with a local reference point. With the data acquired for each position (input signals), the acceleration time series was analysed after applying a low and high filter and a Hanning time domain filter, using GEODAS, the strong motion detection software. By using the obtained acceleration time series and performing the square root of the sum of the squares of each step, the peak velocity values were obtained.

As shown in Figure 10.3, in an overall assessment, at Point A, the mobile seismograph presents higher acceleration values when compared with the fixed seismograph. This position is not part of the building but is near the road and consequently closer to the vibration source.

In the monastery, it is clear that the vibrations measured at the base by the fixed seismograph are always higher than in other points. This position should be considered for monitoring the long-term effect of the traffic-induced vibration analysis.

Another important observation from this preliminary study is the significant difference between the accelerations measured at Point A, near the road and at the fixed point (Point O). This difference shows that there is a significant attenuation in the accelerations measured between the two points. To investigate this attenuation, a new study was conducted to characterise the reduction of the vibration level along the pavement level (see Figure 10.4).

This assessment was performed by employing a two-uniaxial accelerometer system, which allowed accelerations in X and Y to be measured.

Accelerometers (PCB Piezotronics, 393B31 model with sensitivity 9.85 V/g bias level 10.9 V) were used in the study to assess the behaviour of the vibrations along the courtyard between the road and the main façade of the Monastery of Batalha. These accelerometers were fixed in pairs with screws on two cubes, thus performing measurements in parallel and perpendicular to the road direction. An acquisition card with eight-channel and 24-bit resolution was used for communication between the computer and the accelerometers.

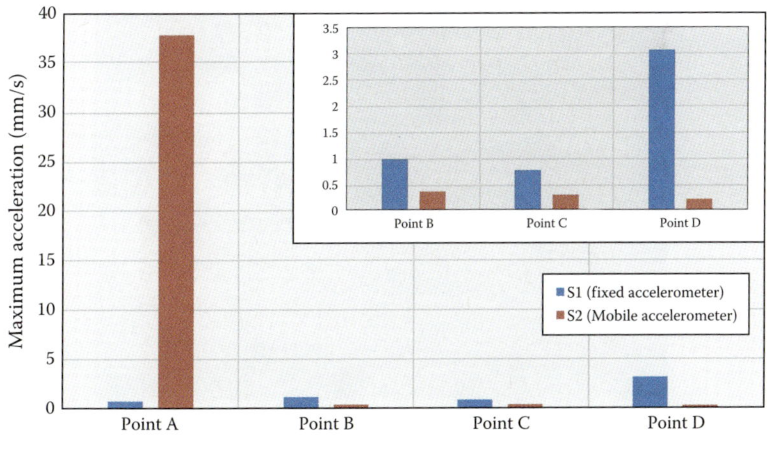

FIGURE 10.3 Acceleration measurements at each point.

FIGURE 10.4 Field test: (a) set-up positions; (b) accelerometers in position 2; (c) accelerometers in position 6.

With the data acquired for each position (input signals), the acceleration time series were analysed after previously applying a low and high filter and a Hanning time domain filter, as described in Section 10.2. A comparison between two points is presented in Figure 10.5, where the huge difference in the measurements of velocity in the X direction between the reference point and Set-up 2 is clear. The velocity measured in the ground near the road is clearly high, greater than 3 mm/s, and in the same alignment near the monument, there is a clear reduction in the measured velocity.

By analysing the results obtained (Figure 10.6), it is possible to verify that the acceleration in the position closer to the road (55.3 m from the façade) is almost 485% greater in Y and 425% greater in X when compared with the acceleration at the reference point near the monastery façade. At the point close to the façade (5.3 m distance), the vibration is 260% greater in Y and 160% greater in X when compared with the acceleration in the reference point near the monastery façade.

With this brief analysis, it can be seen that from the road to the façade of the Monastery of Batalha, the vibrations are strongly attenuated, thus significantly

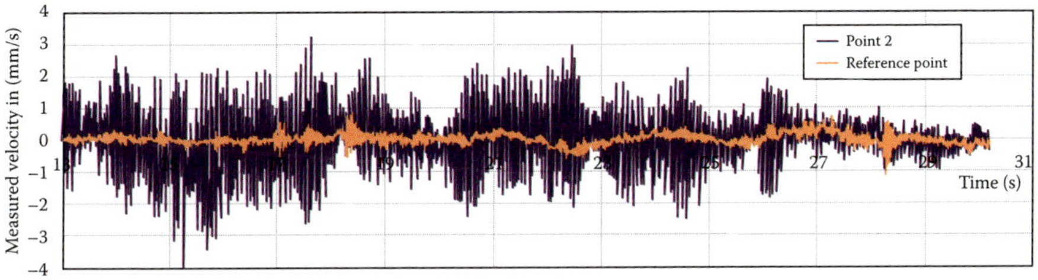

FIGURE 10.5 Example of the velocity measurements in the X direction between the reference point and the second position.

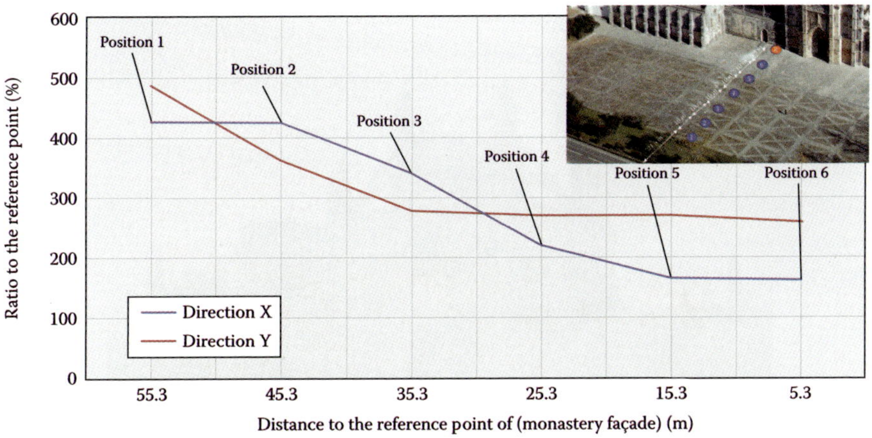

FIGURE 10.6 Ratio of acceleration in each set-up and the reference point in the Monastery façade.

reducing their effect in the monument. Such attenuation is essentially related to distance and soil properties (BS7385-1:1990 1990).

10.3 BUILDING VIBRATIONS AND TRAFFIC-INDUCED GROUND MOTION

Based on the previous study and for the analysis of vibrations in the monastery, a seismograph was placed at two different points on the building façade over several days. The goal was to minimise errors caused by possible local effects at the base of the structure. The points chosen were the Museum of Combatants (I) and the Founder's Chapel (II), shown in Figure 10.7.

In total, 7 days of measurements were made, which resulted in 382 events. These measurements consist of 60 s samples, recorded every 15 min, which amounts to four measurements per hour.

By analysing these samples, it is possible to determine the vibration speeds at any time of the day, thus providing information about the time intervals that are more critical.

Figure 10.8 shows an example of the ground acceleration time histories measured at the point at the Founder's Chapel, the power spectrum, the derived velocity in the three directions and also the vector of the sum of the velocities measured in the three directions.

As can be seen, most ground motion energy lies around 10 Hz, implying that ground vibrations are mainly induced by vehicle wheel-hopping, since the wheel-hopping frequencies for most vehicles are between 10 and 20 Hz (Cebon 1985).

With all the data acquired for each position (input signals), the acceleration time series was analysed after the application of a low and a high filter and a Hanning time domain filter and integrated in time to obtain the results in terms of velocity. Finally, for each signal, the square root of the sum of the squares was performed to compute the maximum value, considering the contribution in each direction. The

FIGURE 10.7 Measurement points.

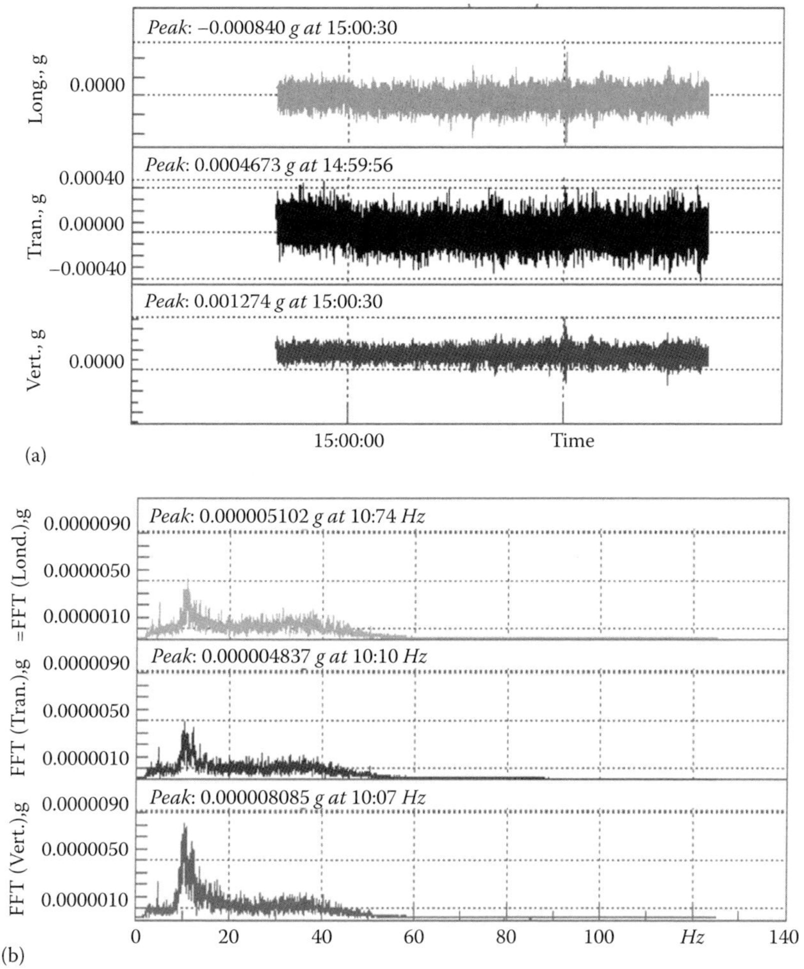

FIGURE 10.8 Example of (a) recorded ground acceleration at the Founder's Chapel and (b) its response spectrum, (c) velocity and (d) sum of the velocities measured in the three directions under current traffic conditions.

conversion from acceleration to velocity aids the comparison of the results with the maximum values proposed by several international standards and guidelines.

The obtained results are presented in Figure 10.9a with reference to the Combatant Museum (225 measurements) and in Figure 10.9b on the Founder's Chapel (157 measurements). The results become more noticeable over time in terms of both amplitude and development.

Based on the obtained results, it can be said that the vibration values in the structure are not greater than a maximum of 2.5 mm/s according to two major international standards: DIN 4150 (German Standards Organization [GSO] 1984) and UNI 9916 (2004). It can also be noted that except for some specific points, most of the values are lower than 1 mm/s (i.e. lower than the maximum allowable value).

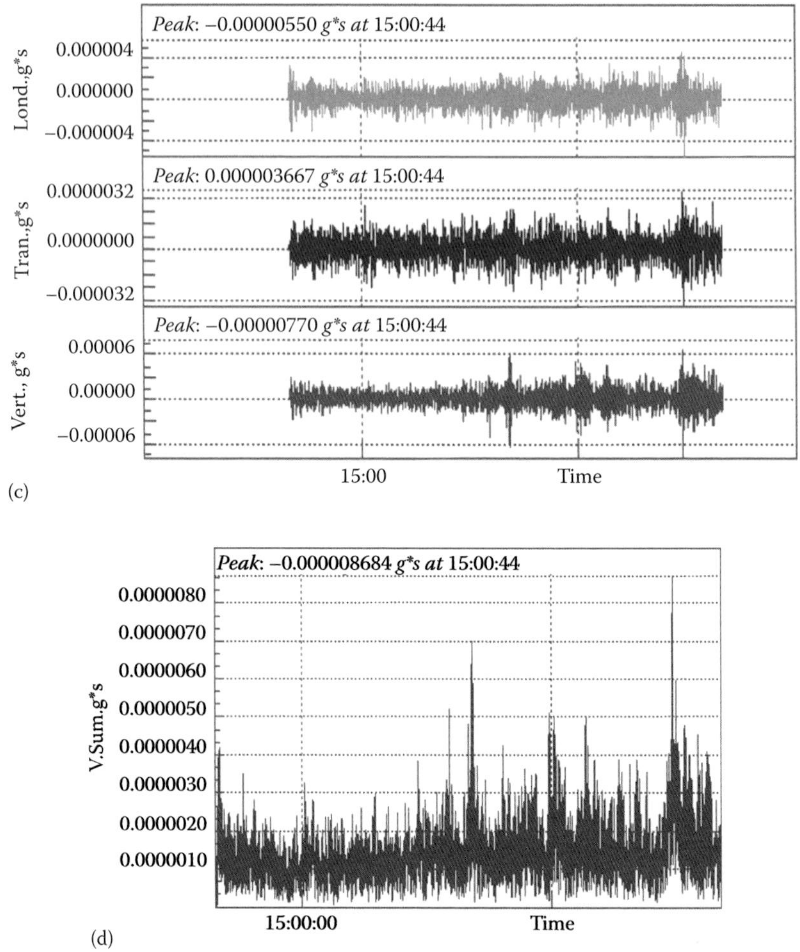

FIGURE 10.8 (Continued) .

10.4 FINAL COMMENTS

Initially, the ground vibrations induced by normal traffic conditions were measured at four sites in the monument. The analysis of the peak values and attenuation was presented and discussed. It was first identified that the base of the front façade of the monument is the most vulnerable point. Furthermore, a strong attenuation was identified between the point near the source (road) and the façade of the monastery.

To characterise this attenuation, an additional study was performed to assess the attenuation throughout the soil in the courtyard between the road and the monastery façade. It is clear that this strong attenuation (around 500%), which is due to the

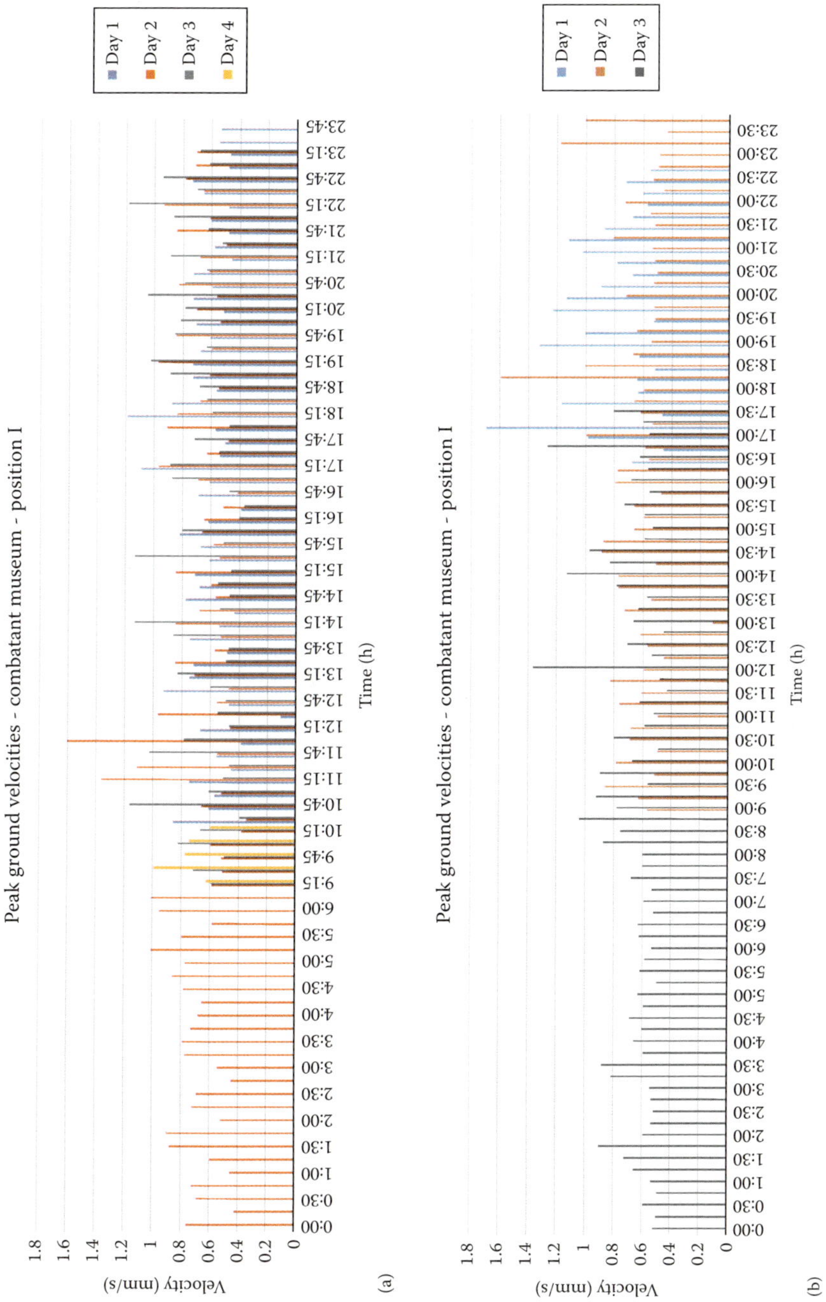

FIGURE 10.9 Peak ground velocities: (a) Combatant Museum (I); (b) Founder's Chapel (II).

distance and soil characteristics, is a key factor in reducing the effect of the vibrations introduced by the traffic.

Finally, using the analysis of continuous measurements performed over several days at the base of the Combatant Museum and the Founder's Chapel, it is concluded that the maximum values of measured velocity do not exceed the values adopted by the German DIN 4150 (1984) and Italian UNI 9916 (2004) standards (2.5 mm/s). In fact, the maximum measured value was 1.69 mm/s. It should be noted that with the evolution of future traffic, it is possible that these values may change, leading to possible structural and nonstructural damage. Retesting is recommended in 5 years' time.

ACKNOWLEDGEMENT

We would like to thank DGPC for allowing us to analyse the Monastery of Batalha, with a special acknowledgement to Dr Joaquim Ruivo and Dr Pedro Redol for all the support given during the field tests in the monastery.

REFERENCES

Antunes, P. F. d. C., H. F. T. Lima, N. J. Alberto, H. Rodrigues, P. M. F. Pinto, J. d. L. Pinto, R. N. Nogueira, H. Varum, A. G. Costa, and P. S. d. B. Andre. 2009. Optical fiber accelerometer system for structural dynamic monitoring. *IEEE Sensors Journal* 9 (11): 1347–1354.

BS7385-1:1990. 1990. Evaluation and measurement of vibration in buildings part 1: guide for measurement of vibrations and evaluation of their effects on buildings. BSI.

Cantieni, R. 2005. Experimental methods used in system identification of civil engineering structures. 1st International Operational Modal Analysis Conference (IOMAC), Copenhagen, Denmark, 26–27 April.

Cebon, D. 1985. The dynamic interaction between wheeled vehicle and road surfaces. PhD thesis, Cambridge University Engineering Department.

Crispino, M. and M. D'Apuzzo. 2001. Measurement and prediction of traffic-induced vibrations in a heritage building. *Journal of Sound and Vibration* 246 (2): 319–335.

D10.4-NIKER. 2012. Guidelines for reliable seismic analysis and knowledge based assessment of buildings: Report, work package 10, NIKER project (2010-2012): New Integrated Knowledge-Based Approaches to the Protection of Cultural Heritage from Earthquake-Induced Risk. Funded by EC under the 7th Framework Program, contract n. ENV2009-1-GA244123.

Ellis, P. 1987. Highway pollution effects of traffic vibration on historic buildings. *Science of the Total Environment* 59: 37–45.

Fernandes, P., M. Vilarigues, L. C. Alves, and R. C. da Silva. 2008. Stained glasses from monastery of Batalha: Non-destructive characterisation of glasses and glass paintings. *Journal of Cultural Heritage* 9, Supplement: e5–e9.

German Standards Organization (GSO). 1984. *Vibrations in Building Construction*. Berlin: DIN 4150.

Gupta, S., W. F. Liu, G. Degrande, G. Lombaert, and W. N. Liu. 2008. Prediction of vibrations induced by underground railway traffic in Beijing. *Journal of Sound and Vibration* 310 (3): 608–630.

Halling, M., I. Muhammad, and K. Womack. 2001. Dynamic field testing for condition assessment of bridge bents. *Journal of Structural Engineering* 127 (2): 161–167.

Hao, H., T. C. Ang, and J. Shen. 2001. Building vibration to traffic-induced ground motion. *Building and Environment* 36 (3): 321–336.
Hunaidi, O. 2000. Traffic vibrations in buildings. Construction Technology Update 39 (National Research Council of Canada).
Johnson, A. P., and W. R. Hannen. 2015. Vibration limits for historic buildings and art collections. *Journal of Preservation Technology* 46 (2–3): 66–74.
Ju, F. D. and M. E. Mimovich. 1988. Experimental diagnosis of fracture damage in structures by the modal frequency method. *Journal of Vibration and Acoustics* 110 (4): 456.
Lima, H. F., R. da Silva Vicente, R. N. Nogueira, I. Abe, P. S. de Brito Andre, C. Fernandes, H. Rodrigues, *et al.* 2008. Structural health monitoring of the church of Santa Casa da Misericórdia of Aveiro using FBG sensors. IEEE Sensors Journal 8 *(7):* 1236–1242.
Oliveira, D. V., L. F. Ramos, P. B. Lourenço, and J. C. A. Roque. 2005. Structural monitoring of the Monastery of Jerónimos. *International Conference on 250th Anniversary of the 1755 Lisbon Earthquake*. Lisbon.
Raguin, V. C. and M. C. Higgins. 2008. *The History of Stained Glass: The Art of Light – Medieval to Contemporary*. Thames & Hudson.
Redol, P. 2003. *O Mosteiro da Batalha e o Vitral em Portugal nos Séculos XV e XVI*. Câmara Municipal da Batalha.
Salawu, O. S. 1997. Detection of structural damage through changes in frequency: A review. *Engineering Structures* 19 (9): 718–723.
UNI. 2004. Criteri di misura e valutazione degli effetti delle vibrazioni sugli edifici (Vol. UNI 9916:2004*)*. Milan, Italy.
Vilarigues, M., P. Redol, A. Machado, P. A. Rodrigues, L. C. Alves, and R. C. da Silva. 2011. Corrosion of 15th and early 16th century stained glass from the monastery of Batalha studied with external ion beam. *Materials Characterization* 62 (2): 211–217.
Xu, Y. L. and X. J. Hong. 2008. Stochastic modelling of traffic-induced building vibration. *Journal of Sound and Vibration* 313 (1–2): 149–170.

11 Rehabilitation of Historical Masonry Buildings in Belarus

Case Studies of Peter the Great's House in Polotsk and Lubcha Castle

Aliaksandr Bakatovich, Nadezda Davydenko, Aliaksandr Ivanenko, Anton Finogenov and Yahor Lazouski

CONTENTS

11.1 INTRODUCTION

The main goal of this chapter is to analyze national experience in the restoration of historical masonry buildings, using the example of national historic landmarks. Two historical monuments, Peter the Great House in Polotsk and Lubcha Castle, were examined with GPR for soil scanning and a Ferroscan device for the detection of metal elements in masonry. The authors also describe nondestructive methods for monitoring the properties of the structural materials and propose how to restore their historical authenticity. The results allowed the researchers to develop a 3D model for further restoration and reconstruction of the named monuments.

11.2 PETER THE GREAT'S HOUSE IN POLOTSK

11.2.1 Historical Reference

Peter the Great's house was built in 1692 on the site of a wooden building in Polotsk. It was partially rebuilt at the end of the eighteenth and the beginning of the nineteenth centuries. The building is in a baroque style and was used as a dwelling house. The architectural style of the house includes ornamental windows, a principal front and flank fronts. The front doors are decorated with carved wooden plants. The building is located in the historical centre of Polotsk in Nizhne-Pokrovskaya Street along the Western Dvina River. This historical house was one of the few buildings in Belarus of urban style and is located in the Polish-Lithuanian Commonwealth.

According to chronicles, Peter the Great stayed in the historical house during 1705. The memorial plate on the house, placed there in 1952, proves this fact. The Northern War (1700–1721) that took place in modern Belarus is the reason why Peter the Great stayed in Polotsk.

During the Great Patriotic War (1941–1945), the house was considerably damaged, but was restored in 1949. During the post-war years, it housed the children's library, named after Leo Tolstoy, and a monument to this great Russian writer stands in front of the house. Since 1998, it has housed the permanent exhibition 'Ramble along Nizhne-Pokrovskaya Street'. It is the first monographic exhibition in Belarus dedicated to the history of just one street. Here, one can learn the history of the street and the interior of some buildings typical of the beginning of the twentieth century, including the way people lived and worked.

Peter the Great's house is on the list of historical and cultural heritages of Belarus and is protected by the government.

The bearing constructions of the building have been worn down and greatly damaged during the 60 year of its operation. A technical examination of the house was carried out in 2007. A plan was outlined to repair and restore Peter the Great's house.

11.2.2 Dimensional and Planning Design of the Building

Peter the Great's house is an irregular rectangular construction, 18470 × 8830 × 18640 × 8210 mm in size. There is a basement in the building. The floor and basement plans are shown in Figure 11.1.

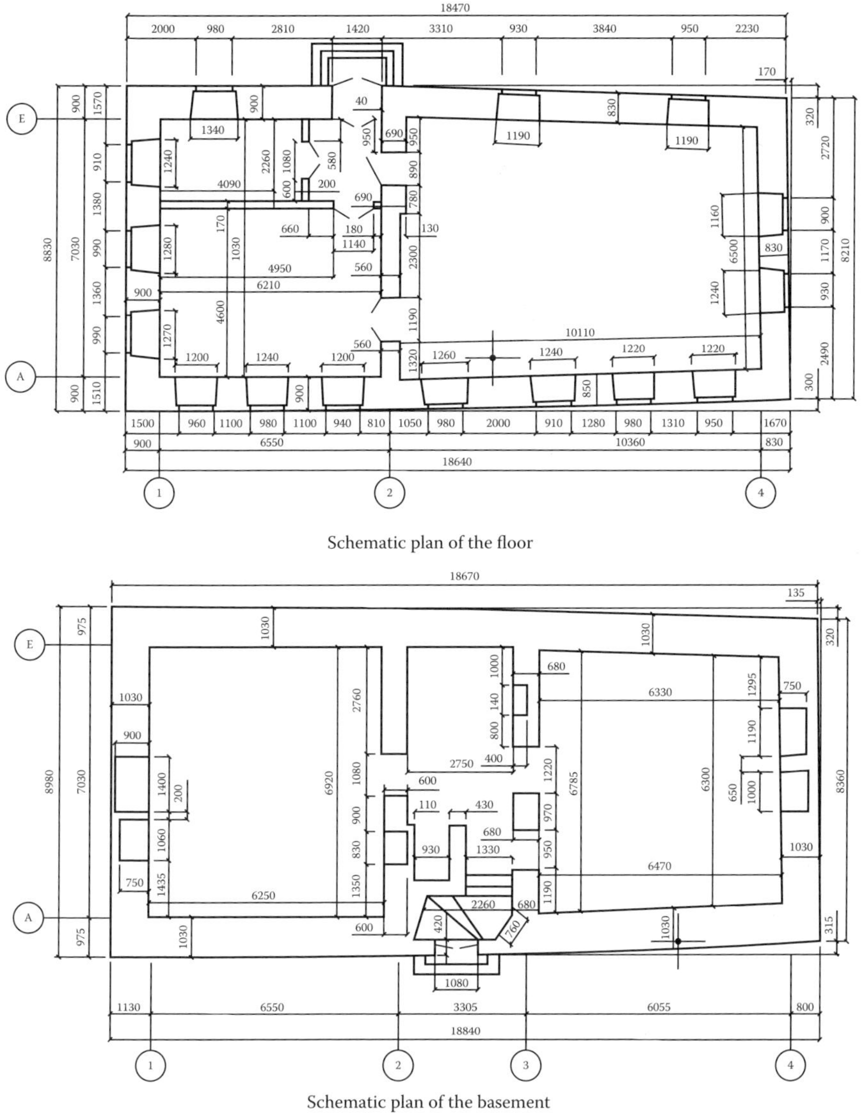

FIGURE 11.1 Schematic plan of the floor and basement of the building. (Scheme made by the authors.)

The external and internal walls are made of ceramic whole brick, calcinated under low temperatures until the desired glassy phase has been achieved. The masonwork was made on the basis of a lime-sand, brownish-cream mortar, and the components ratio was 1:2. The clay substance in the masonry mortar is in the form of admixture in which sand is used, which determines the brown colour of the mortar.

The ground floor walls have been boarded and lath plastered with lime and sand mortar.

The ceilings between the ground floor and the basement are brick vaulted with bearings on the exterior walls endwise '1'–'2' and '3'–'4' (arch width is 6.52–6.6 m, top of the arch is 2.55–2.69 m above floor level) and the exterior and the interior walls endwise '2'–'3' (arch width is 2.75 × 4.5 m, top of the arch width is 2.3 m above floor level).

The garret floor is made of timber: a bar section of 180 h × 180 mm and a step of 1500 ± 50 mm. The heat insulation of the garret floor is made by filling the cracks (160–180 mm tick) with slag.

The roof is a span roof, which means that the roof has two slopes of equal breadth and pitch, meeting at a common ridge. The measurements of the wooden rafters are a section of 190 h × 40 mm, making 1500 ± 50 mm. The top tie is 190 h × 40 mm at a height of 1260–1360 mm, the lower tie equals 200 h × 200 mm. The step of roof boarding, 70 h × 70 mm, is 230 mm. Wall plates are 150 h × 200 mm.

The floor of the ground floor is supported by wooden liggers. There is no floor in the basement, only soil.

The beds are made of ceramic whole brick on a lime-sand mortar using quarry stone. At the entrance, the depth of the bed is 2.2 m above the ground floor.

11.2.3 Reparation and Restoration Works Based on Technical Examination Results

11.2.3.1 Technical Examination

The construction of this historical building has been the same for the last 300 years, at least regarding specific climatic conditions such as temperature oscillation and the quality and type of precipitation. Climatic conditions such as alternative freezing and defrosting of the structural materials during autumn, winter and spring, which are typical of this region, have a tendency to influence the construction. Under these conditions, any operation concerning the construction of walls can be seriously impaired. A technical analysis of Peter the Great's house was made according to instructions provided by the National Polotsk Historical Preservation Society.

The bearing exterior and the interior brick walls are in good conditions. Plaster has been destroyed endwise 'A' axis on the socle wall of the front exterior wall (Figure 11.2). As a result of the alternative freezing and defrosting process, plaster has also been destroyed on the drip cap. A certain amount of paint flaking has also been observed. Figure 11.2 also shows plaster destroyed endwise'4' on the socle wall of the front wall as a result of freezing and defrosting.

An analysis of the exterior wall endwise 'B' showed the destruction of plaster on the socle wall and the drip cap, some paint flaking and mould on the finishing surface in the lower part of the wall. There is a horizontal crack at the crossing of axes 'B' and '1'.

An analysis of the wall endwise '1' found that the destruction of plaster was as a result of defrosting and the lack of a drip cap at the wind opening.

The Ferroscan Hilti PS-200 was used to find metal tension bars that strengthened the brickwork, making it possible to avoid vertical cracks appearing in the walls due to the different settlements in the foundation during the initial use of the building.

Plaster destruction of socle wall endwise "A"

Destruction of the basement exterior wall endwise "4"

Destrucion of plaster and masonry mortar of the upper part endwise "3" – "4"

Cracks in joint points of the ceiling to exterior wall endwise "A"

Rotting and mould of the rafter in bearer endwise "B"

Destruction of interior part of masonwork endwise "B" and "4"

FIGURE 11.2 Typical damage to the constructions. (Photo taken by the authors.)

The results of the scanning showed metal tension bars throughout the building at least on the exterior walls above the windows. The metal tension bar that was embedded in the brickwork between the inner rows of brick had the shape of forged metal rows 1.5–2 m long and a section of 35–40 h × 5–6 mm. Connecting joints were forged at the end of each row. The results achieved were a full-length solid tension bar made by joining separate rows with tags overlaying the adjacent rows one above the other and then fixing them with metal pivot bolts (Bakatovich et al., 2007).

The brick walls from the basement side are in satisfactory condition. However, some masonwork has a crack measuring 70 mm deep, which is caused by the alternative freezing and defrosting process carried out in a moist state. This occurred because of the lack of horizontal and vertical damp proofing and the fact that there is no heating in the basement during winter time.

The arched floors between the ground floor and the basement are in a satisfactory condition. There is a destruction of the brick masonry in the arched floors endwise '1'–'2' and '3'–'4' 50 mm deep, and a destruction of plaster endwise '2'–'3'. Destructions were caused by alternately freezing and defrosting in water-saturated state.

The shock-and-pulse device, IPS MG 4.03, and basement samples used compressive strength and masonry mortar in the blockwork, arched floors and the basement. It was concluded that ceramic solid bricks have a compressive strength of 9–11 MPa, and the malm bricks strength is 0.8–1.1 MPa.

The arch in the wind porch of the basement is in the shape of brickwork that has been laid above a metal-shaped frame. The metal parts of the frame are in poor condition covered by a thick layer of corrosion.

A technical analysis of the wooden bars of the garret floor showed considerable damage. Cracks at the joint points of the floor and the walls have been caused by insufficient rigidity of the wooden bars. There is also a large amount of rotting on one of the bars endwise '1'–'2'.

Rotting of the roof frame boards (about 70%) and the top of the rafters (about 60%) was caused by drenching, and traces of mould were observed. There is also rotting of five rafters of the reference nods endwise 'B', sagging of two lower ties, and a lack of a wooden part on the rafter endwise 'A'.

The nondestructive method (GOST 16483.3, GOST 16483.10) was used to find the compressive and bending strength of the wooden elements of the garret floor and the framed roof. The compression strength of the elements found in the garret floor and those of the frame roof made up 47 MPa (compression along fibres), and the bending strength was 38 MPa.

The metal plates used to cover the roof are not worth using. Seeping of the roof covering was caused by poor lock joining of the roof plates, which in turn led to the drenching and rotting of the wooden parts of the frame.

The brick bed is in poor condition. There is some destruction of the exterior and the interior parts of the masonwork, 150 mm deep, caused by alternative freezing and defrosting in water when in a saturated state. There is also a lack of vertical and horizontal water proofing.

Typical damage to the construction is shown in Figure 11.2.

11.2.3.2 Repair and Restoration Works

Reparation and restoration work was carried out to strengthen the bearing constructions of Peter the Great's house in Polotsk.

11.2.3.2.1 Restoration of Damaged Bricks in Masonry Arches and Walls

Support stands were installed, covering the whole surface of the arches, to provide the restoration of the brick arches in the basement. The more seriously damaged bricks (more than half-destroyed) were pulled out of the masonwork and replaced with new bricks. Masonry mortar was used to fix the new bricks. The composition of the mortar is lime (0.8), cement (not more than 0.2) and average grained sand (3). The surface was cleaned of remnants and dust left behind by the mason work. All was removed before starting the process of covering the surface with new mortar. Lime, type I, and washed sand without any powder or even clay admixtures were used in the masonwork mortar (Sinavski and Usoskai, 2008).

Parts of the less destroyed bricks (up to 20 mm deep) were filled with 'FuncosilRestauriermoertel' mortar (manufactured by Remmers, Germany), which is made of cleanraw mineral materials. The mortar colour was chosen during the restoration works.

To restore destroyed bricks (more than 20 mm deep), the raising of plug was implemented and two layers of priming mortar 'FuncosilGrundiermoertel' were applied. The restored parts were treated with compressed air and well dried after priming. Then, a thin mortar FuncosilGrundiermoertel (its wet variant) was used. Sections of the brick joints were preserved in order to fill them later with masonry mortar. The surface of the restored brick was cleaned with a microporous rubber.

11.2.3.2.2 Filling New Mortar into Brick Arches and Walls

Destroyed masonry mortar was mechanically removed using metal hairbrushes and electric tools. The joints were cleaned of dust and remnants of mortar after mechanical treatment. The damaged joint was opened up to 20 mm deep, and the joint base was washed. The joints were filled twice with 'FuncosilFugenmoertel' while pressing down hard on the joints. Subsequently, the surface was levelled with a putty knife.

Hydrophobisation of the masonry surface of the arches and walls was carried out with a water-repellent soak 'FuncosilFassadencreme'. A special soak was used in the preservation of the bricks and mortar. All this occurred four weeks after the masonry restoration work. The surface was covered once. The original white colour of the hydrophobisation layer disappeared 15–20 min after coating.

Strengthening of the brick base was carried out to enable use of the building, as shown in Figure 11.3 (Bakatovich et al., 2007).

Following a close analysis of the foundation soils in the old area of Polotsk, it was found that the foundations of the previous building did actually exist, and that the abandoned underground lines threatened nearby buildings and constructions. Specialists of the Polotsk State University used georadarProEx (Sweden) and antennae 250 MHz on Peter the Great's house and the adjacent territory (Kremnev et al., 2008).

As a result, residues of substructures, the base of a merchant's storehouses and a drain box were found about 25 m away from Peter the Great's house. An interpretation of the results of the georadar shape that lay above the curved arch and residues of the drain box are shown in Figure 11.4. An analysis of the results showed no threat to the storehouses or the drain box residues of Peter the Great's house. The GPR results (search for residues of underground structures of a merchant's storehouses and a drain box) were proved during pit works on a new office building.

11.2.3.2.3 Horizontal and Vertical Waterproofing of Walls

A set of works was carried out to prevent moisture saturation of the walls. Surface levelling of the land according to the layout of adjacent land was carried out, as well as the installation of back reinforced concrete walls endwise 'B' with artificial stone trimmings. A paving concrete tile riprap (700 mm thick) was installed around the building.

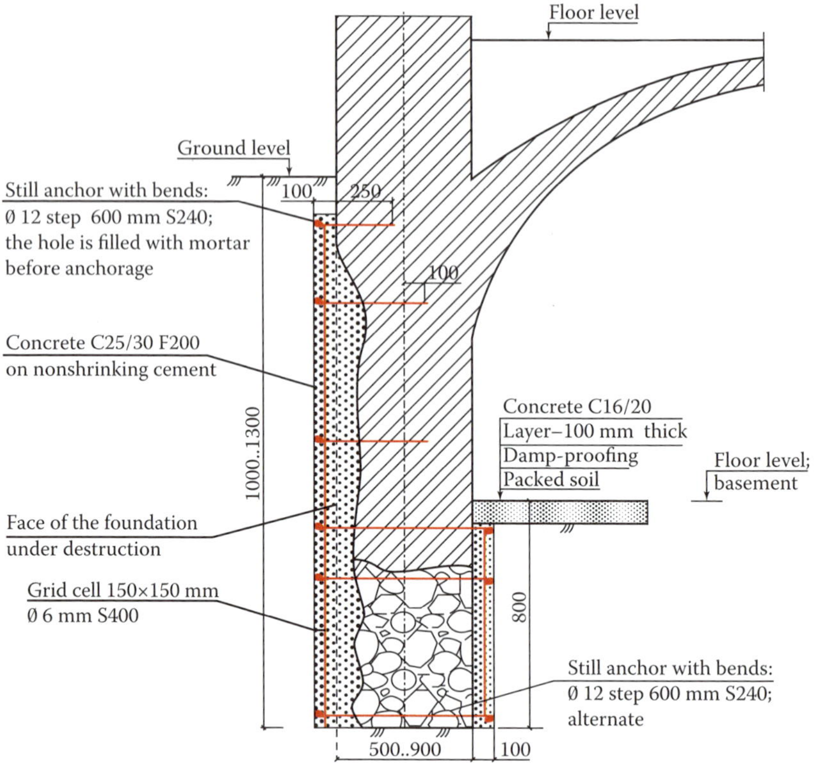

FIGURE 11.3 Scheme for strengthening the brick base. (Scheme made by the authors.)

To protect the walls from capillary moisture suction and from a saltpetre appearance, which would destroy the masonry and plaster, horizontal damp proofing was carried out by means of alternative chemical composition injections on the level of the basement wall and the lower wall part of the beam ceiling above the basement.

The 'Elakvablok' composition (manufactured by Baumit-Bel, Belarus) was used as the waterproof injection material. The exterior wall was injected at the level of the basement wall floor from inside through 16 mm holes and a 150 mm step. The angle of the holes was 20°–30°, the depth was two-thirds the thickness in two lines. The second line of holes was drilled at a height of 100 mm and had an alternative shift. The waterproof composition was poured into the holes. A moisture cut-off was applied on the interior walls, mainly at wall level and the basement floor, but on the exterior walls, the moisture cut-off was applied on the lower part of the beam ceiling. The cut-offs were made by injecting through the holes that were drilled from both sides of the walls.

The waterproof composition `Polimiks-GS' (manufactured by PK Radeks, Belarus) was used to vertically waterproof the bed foundation to the land level, and then the walls, 0.5 m high.

Air ducts and heating plants were installed to dry (sanitation) the wall surface, providing a proper microclimate in the basement.

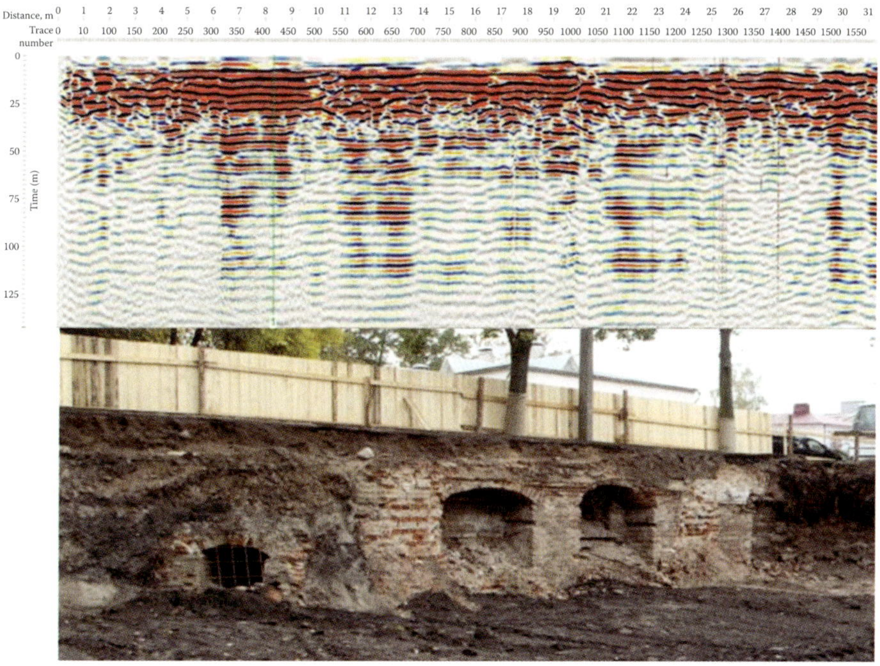

FIGURE 11.4 Results of georadar shape of underground parts of merchant's storehouses and a box drain. (Photo taken by the authors.)

11.2.3.2.3.1 Restoration of Damaged Plaster The sections where the plaster was destroyed were mechanically removed with rough metal scrubbers, scrapers and warm washing. A water drainage system was installed to drain and clean water from the socle part of the building. The plaster was removed step by step.

Sections where the damage (flaking, peeling) was not too severe were removed by regrating the plaster after preliminary digging. The destroyed top lift was removed by means of metal scrapers; the surface was watered beforehand and ground with a lime and sand composition. Simultaneously, the cracks were pointed out and filled with lime mortar containing marble and rock powder and subsequently rubbed with float.

The open plaster cracks were levelled with the adjacent surfaces of a mortar (ratio): lime (0.5 ÷ 0.7), cement (0.3 ÷ 0.5) and sand (3). Concentrations of CaO and MgO not less than 65% and cement M500 were used in the lime for the mortar. Dust and remnants of the damaged plaster were removed from the surface before plastering (Sinavski and Usoskai, 2008).

Colonies of biodestructors (fungi, mould) that were destroying the surface plaster were also removed before conservation works began. Solutions of the biocide preparations 'BaumitSanierLosung' (manufactured by Baumit-Bel, Belarus) and 'HistolitAlgen-Entferner' (manufactured by Caparol, Germany) were used for the microbiological treatment of the surface. Molar brushes were used to cover the wall surfaces twice with an antirot substance. The intervals between treatments were not less than 12 h. Additionally, the air temperature for the wall surface treatment was not less than +15°C and humidity was 40%–70%.

To strengthen the old plaster and even out the level of absorbency in the new and old plaster, the wall surfaces were coated with a special primer paint. Paint compositions were applied on the walls not less than 4 weeks after plastering.

11.2.3.2.3.2 Other Repair and Restoration Works A waterproofed concrete floor, 100 mm thick, was laid on the basement soil. Concrete C25/30 was applied.

The metal frame of the wind porch ceiling was considerably damaged. So, the ceiling was dismantled. Installation of a reinforced concrete wind porch ceiling (concrete C30/37 used) was then followed by the framework installation.

Test calculations on the geometry of the garret floor beams and parts of the rafters showed that they were not strong enough to bear the actual loads. Heat calculations showed that the heat resistance of the garret floor was only 1.23 $m^2 \times {}^\circ C/watt$ (Bakatovich et al., 2007). A garret floor ceiling and new rafters were installed on wooden beams. The garret floor was heated up with mineral wool boards 400 mm thick. The ventilation pipe of the clinkered brick was then roof tiled. The roof was covered with steel plates coated with polymeric dark-green paint as well as a roof drainage system, as shown in Figure 11.5.

11.3 LUBCHA CASTLE

11.3.1 Architecture of Lubcha Castle Complex in Historical Context

The Lubcha Castle complex is situated on the Neman River in Belarus. Several buildings are located in the castle grounds: entrance and south towers, a palace, a wing and an outdoor dwelling. The rehabilitation and restoration works are being carried out under the patronage of a charitable foundation 'Lubcha Castle' and its head Ivan Pechinsky.

FIGURE 11.5 General view of the building before and after restoration. (Photos taken by the authors.)

Lubcha Castle follows all the best Renaissance traditions, although not as strong as the traditions in other European countries. Counter-Reformation ideas were typical of the time. The castle reflects the typical architectural features of the time: quadrangle loopholes in the towers' corners, almost causing a square fortress effect.

The complex is situated on a high hill, built with boulder masonry, on the left side of the Neman. Its territory is 85 × 80 m. The castle is surrounded by a moat which runs into a river and could be filled with water. In the corners of the castle were four towers. Unfortunately, only two still stand today. These towers had defensive purposes (Tkachov, 1978).

To access the territory where the castle is situated, you need to go through the entrance tower (Figure 11.6). At the end of the 1920s, an earth fill replaced the former wooden bridge, which leads to the castle.

A square tower is situated in the southern corner of the castle site (Figure 11.7). Having analysed the brick and mortar, one can dwell on the simultaneous erection

FIGURE 11.6 General view of the entrance tower before and after its restoration (2008–2016). (Photo taken by the authors.)

FIGURE 11.7 General view of the south tower before and after its restoration (2008–2010). (Photo taken by the authors.)

of the remaining constructions – the entrance and the south towers. Unfortunately, neither the wooden east tower nor the brick north tower was preserved.

When considering the possible dates for the beginning of the construction works, the last third of the sixteenth century seems most probable. During the construction works in 1968, a vane with a coat of arms and the year 1581 forged on it was found on the castle's territory. The date, stamped on the vane, has been proved through archaeological studies (Chernavski and Saitgarzeva, 1984).

The castle complex was extensively damaged during the Great Northern War in 1655. The castle was damaged for a second time when Russian Cossacks occupied Lubcha. Inventories of 1661 and 1662 have proved the existence of three corner towers in the castle. According to the inventories, the south and the entrance towers were connected by a stone fence with a wooden gallery. The space between the entrance and the north towers was built with a solid line of stone constructions, which had been used as living and utility rooms.

On August 21, 1730, Lubcha was set on fire. It is possible to ascertain that great reconstruction works were carried out on the tower after the fire; at least this is the perception as new buildings are mentioned in descriptions of the castle from 1738. A smart palace was seen as the most impressive among the buildings. The building, measuring 78 × 24 ells (50.7 × 15.5 m), had spacious rooms and tiled stoves decorated the palace.

At the beginning of the 1860s, construction work began on the stone palace in the north-east line of the castle territory. The newly built palace was symmetric with a two-storey cubic content in the centre and two adjacent one-storey outbuildings. Decorative towers with merlons were situated in the corners of the facades. The facades were decorated with eaves and avant-corps. Front facades had merlons and gables with rectangular holes imitating loopholes. In the centre of the composition was a balcony that could be seen from the yard. The balcony was 8 m under hung, leaning on the columns. Window openings were decorated with canopies, and the doorways were trimmed. The rooms in the centre of the palace had a corridor system. There were sitting rooms, halls and living rooms. Utility areas were inside the outhouses. Moulded elements and stucco mouldings decorated the interior of the palace. The interior of the palace, dating from 1876, proves that the palace was made out of brick and it was shingle-roofed: its dimensions were length of 44, width of 18 and height of 13 arshins. The outbuilding's length was 15, its width was 15 and its height was 8¾ arshins.

No wing appears in the drawing of 1869; it was probably dismantled in the 1840s or 1850s. A new castle owner, Eduard von Falz-Fein, began the construction of a new brick wing at the end of the 1890s. The building, as well as the palace, had small towers in its corners. In 1915, during World War I, Lubcha Castle was in the line of German defence and military operations greatly damaged it.

In the 1920s, a wooden outhouse was built onto the living wing of the castle. To ease the difference between the construction periods, two brick towers with merlons were erected in the corners of the outhouse. The wing and the outhouse remain today (Barkun, 1983).

Lubcha secondary school was built in 1947–1949 on the foundations of the remaining walls (after dismantling the upper and part of the ground floor) and a total replanning was undertaken. Until recently, there had been a library. Volunteers, helping to restore the castle, live here in the summer.

11.3.2 Technical Survey of Castle Structures and Their Repair and Restoration Works

11.3.2.1 Castle Towers and Defensive Walls

The entrance tower is a massive cube with an octagonal prism at a height of 9.5 m (Figure 11.6). The tower foundation is 9.93 × 9.97 m and its height is 21.2 m. Three tiers make up the defence of the tower. Six loopholes on the first tier of the first floor are made for firearms, and there are seven loopholes on the second tier. There are two round musket loopholes in each wall under the roof. The bed of the castle, which is 2 m deep, is made of huge broken fieldstone bricks on a lime mortar.

A technical analysis of the castle's building was carried out in July 2008. The material used in the castle's construction included ceramic bricks and lime-sand mortar. The brick size is 310–320 mm long, 150–165 mm wide and 65–80 mm high. The mass of a unit is 5.8–6.9 kg. The strength of the ceramic bricks and the brickwork mortar of the entrance tower and other buildings of Lubcha Castle was tested on selected specimens with a shock-and-impulse device IPS MG 4.03.

The compressive strength of a ceramic brick is 10.1–12.3 MPa and its bonding strength is 3.5–4.3 MPa. The bricks are homogeneous with some small cells. This proves the high quality of the ceramic bricks and the moulding technology.

The masonry mortar colour is light grey-yellow. Silica sand with grains of up to 5 mm is used as a filling aggregate. The thickness of the vertical and the horizontal joint of the brickwork varies from 5 to 30 mm. There are clots of unmixed lime, 10 mm, in the mortar. The compressive strength of the lime-sand mortar in the samples taken was 1.4–1.7 MPa.

Alternative freezing and defrosting in a water solution caused the destruction of the plaster and the front layers of the brick. Huge parts of the brickwork are damaged, at least at the junction points of the cube to the octagon. The upper layer of the brickwork was destroyed to a depth of 350 mm (Figure 11.6). The building analysis showed partial destruction of the plaster and the inner brickwork of the walls.

There is a through-wall crack, passing through the walls and the arch of the first floor at the top of the arch above the entrance. There are other vertical cracks in the tower walls. Local dwellers stated that the cracks appeared a long time ago. Special sensors were used to control the development of the crack width (Bakatovich et al., 2008).

A reduction in the earth fill thickness of the moat bed was due to soil creeping, which is the reason why the cracks appeared. Schemes of the characteristic damage to the exterior walls of the entrance tower facades are shown in Figure 11.8.

The tower walls were scanned with Ferroscan Hilti PS-20 with the objective of finding out if it was necessary to strengthen the tower. As a result, metal tension bars were found in the exterior walls above the entrance arch of the first floor. It is probable that the tower architects used the bars to improve the brickwork's strength and prevent the development of vertical cracks because of the uneven shrinkage of the ground. The metal bars did not prevent the appearance of vertical cracks but they

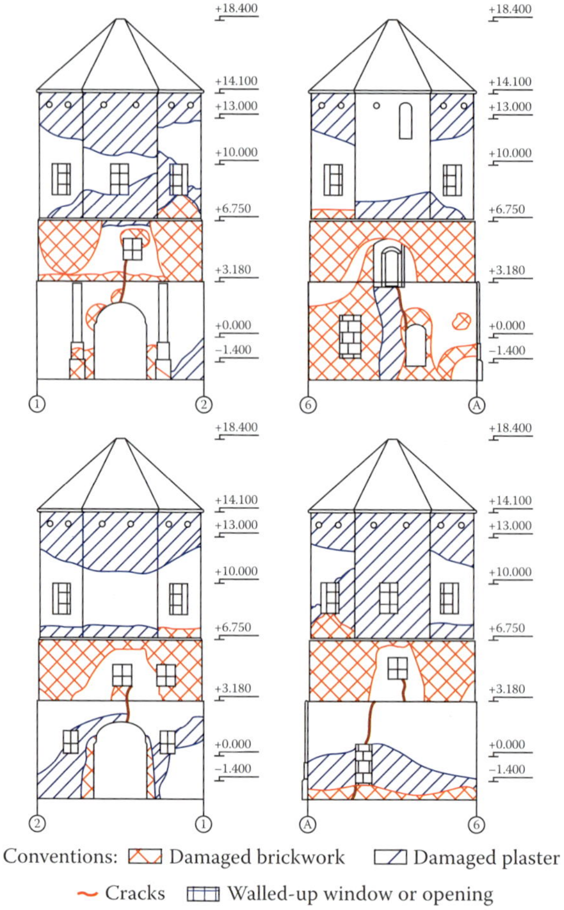

FIGURE 11.8 Schemes of the characteristic damage to the walls on the facades of the entrance tower. (Schemes made by the authors.)

did save the tower from irreversible damage caused by the unfavourable ground base (slope position). Ferroscan results proved that the bars were made up of separate metal strips up to 2.5 m long when joined together.

In July 2009 and August 2011, on checking the sensors no flaws were detected. Thus, we can say that there is a need to stabilise the crack opening and strengthen the walls.

During the analysis of the outer plaster of the tower walls, coloured layers were found. Decorative plaster sgraffito (a white layer was found under a black layer) was used in the original plastering of the tower.

Reparation and restoration work has been carried out over the last eight years. In 2008, wooden floor beams above the second floor were installed in the tower. A rough wooden floor was built on the beams in the winter of 2009. Peripheral scaffolds were installed in the summer of 2010 and 2011. The destroyed parts of the masonwork and the plaster at the front were restored in the summer periods of

2012–2014. In 2014, temporary scaffolds were dismantled and a dome with a roofing sheet platting was mounted. Its architectural authenticity was preserved. Parts of the roof covering, made from a ceramic tile, were laid and the wooden window frames were replaced. Since 2105, decorative plastering works for black-and-white sgraffito along the ledges of the window openings on the second floor and the tower entrance trimming have been carried out (Figure 11.8). Completion of the works was planned for the autumn of 2017.

Repair works in the tower were carried out simultaneously. The floor construction of the second floor above the brick arch was dismantled. The construction of the bed for a new boarded floor was also carried out. Lost elements of the plaster found inside the tower were restored. The installation of wooden gates with decorative forging is planned for the last stage of the works.

11.3.2.1.1 South Tower

The other remaining tower – the south tower – has three storeys and a basement (Figure 11.7). The bed of the tower is 8.06 × 8.35 m. The wall thickness is from 1.56 m in the basement to 1 m in the second floor.

According to the inventory of 1601, there were 4 loopholes for rifles on the ground floor, 4 for guns on the first floor and 12 for rifles on the second floor. A cellar with an iron-barred, wooden door on hinges was found in the socle part of the tower. A coffin chamber with iron-clad doors was situated on the ground floor. There were latticed windows.

The ceramic bricks used in the tower construction had the following size: length of 305–315 mm, width of 140–160 mm and height of 70–80 mm. The weight of a unit is 6.0–7.1 kg. The compressive strength is 8.9–10.2 MPa and the bonding strength is 3.2–3.8 MPa.

The mortar samples tested had a compressive strength of 1.8–2.4 MPa. There were no homogenous parts that had mixed lime elements.

Due to the absence of roofing, the brickwork of the arch between the ground and the first floor has been badly damaged by freezing and defrosting. The brickwork is destroyed to two-thirds of the brick thickness, and a salt coat covers the surface. The arched floor is very poor. The arch is unloaded by the first floor load then transferred via wooden beams to the walls.

Considerable damage to the towers walls has been found. Partial destruction of the bricks and the mortar of the interior half layer of the bricks was found on the socle of the southern and eastern wall. This destruction was caused by alternative freezing and defrosting due to water saturation (Figure 11.7). There are cracks on the north and south walls of the tower. The uneven foundation settlement is the probable cause for the appearance of defects. The crack opening is no more than 3.5 mm. In the summer of 2008, sensors were installed on the walls to study the crack opening process. At the end of 2010, the sensors had not registered any increase in the crack openings. Thus, we consider that the bed has stabilised (Bakatovich et al., 2008).

Schemes of the characteristic damage of the exterior walls of the south tower facades are shown in Figure 11.9. The exterior walls of the south tower were scanned by Ferroscan to find metal tension bars in the brickwork. The use of the Hilti PS-200 for scanning found metal tension bars in the ground and first floors of the south

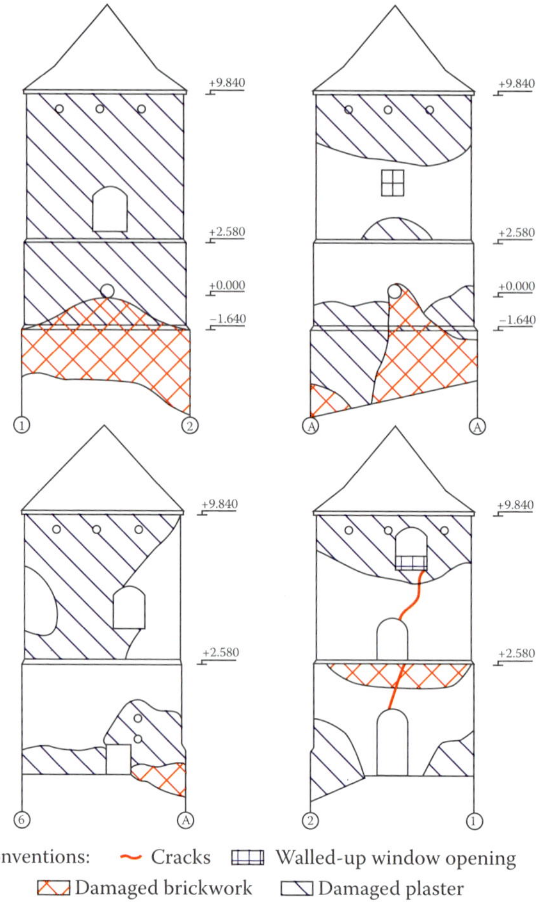

FIGURE 11.9 Schemes of the characteristic damage to the walls on the facades of the south tower. (Schemes made by the authors.)

tower. Researches have proved that the same method of bar making used on the south tower was used on the entrance tower. The bars were fixed with pintles with holes at their ends. Metal tension bars prevented the development of vertical cracks opening up on the exterior walls of the ground and first floors, thereby stabilising the opening process. The use of metal tension bars in the brickwork prevented the tower from total destruction up to the present time, despite its location on the slope.

The main reconstruction works of the south tower were carried out in the spring of 2007–2011. A wooden roof frame was installed in the eaves of the restored brickwork, and the frame was covered entirely in ceramic tile. A vane, similar to that found in 1968, was placed on the roof. A rotten floor between the first and second floor was replaced, as well as in the attic. A stairs-and-observation area next to the north tower was installed. The outer walls of the tower were plastered and painted (Figure 11.7). Decorative plaster covers the eaves – black-and-white sgraffito. The sgraffito picture was imitated using the remaining pictures that had been left behind.

Inner finishing works were carried out in the south tower in 2012–2014, and authenticity was preserved. Restoration of the arch brickwork was carried out in the summer of 2016. The most damaged bricks were replaced with whole bricks of the same size on a lime and cement mortar. A decision to use brickwork to conserve the arch or to back-plaster the arch with chemical anchors is yet to be made.

11.3.2.1.2 North and East Towers

Georadar (GPR) sensing of the base location was carried out on the supposed area of the north tower location with the objective of finding out its real coordinates. The GeoradarProEx (Sweden) was used for sensing. All works were carried out before the archaeological excavations.

Excavations on the probable place of the north tower were carried out in July 2016. The brickwork of the tower bed was found to be 2 m deep of the castle's area (Figure 11.10). Both the inside and outside layers are made from whole ceramic brick on lime-sand mortar. Rubble and broken bricks on the mortar were used for masonry purposes. The distance between the facets of the eastern and the western walls of the bed was 9.8 m.

The compressive strength of the brick samples tested was 9.3–11.2 MPa and the bonding strength was 2.8–3.5 MPa. The compressive strength of the masonry mortar tested was 1.4–2.1 MPa. The mortar composition, the brick sizes and the strengthen characteristics prove the period in which the towers were built.

The absence of a bed under the northern tower wall, facing the Neman, was fixed. One of the reasons for the absence of a wall is the destruction of the masonwork that occurred during the wall improvement of the Neman shore at the beginning of the twentieth century.

Design for the tower restoration is under consideration. Lost parts of the bed will be restored. A solid reinforced belt along the masonwork is required to offer proper restoration works to the new and old part of the bed. The architecture of the north tower is similar to that of the entrance tower, the only difference being the dome.

FIGURE 11.10 North tower bed (2016). (Photo taken by the authors.)

The search for the east tower bed with a GPRProExsensing had no success. The ditch works – mainly on the pits – also came up with no results. However, a drainage system of ceramic pipes from the castle square to the river was located. The east tower was made up of logs and was soon destroyed during military operations. The bed of the logged tower was not strong enough to withstand a military incursion. So, the bed could have gone to ruins over the last three centuries. Periodical replanning of the castle territory could also be one of the reasons.

11.3.2.2 Wing and Outbuilding

A two-storey wing and a one-storey outbuilding are now used as residential buildings (Figure 11.11). The size of the wing is 10.4 × 12.1 m and its height is 8.7 m. The size of the wooden outbuilding is 10.3 × 12.1 m and its height is 5.6 m.

The compressive strength of the sampled ceramic brick was 9.5–11.3 MPa and the bonding strength was 3.6–4.5 MPa. The brick sizes were 260–270 mm in length, 120–130 mm in width and 70–75 mm in height. The weight of a unit was 3.9–4.1 kg. The sizes and the physical and mechanical characteristics of the brick used in the wing are similar to those of modern whole ceramic bricks. In order to make the wing more impressive, fashioned ceramic brick was used to decorate the towers of the wing. The compressive strength of a lime-cement masonwork mortar varied from 2 to 2.5 MPa.

Masonwork on the socle part of the wall was damaged by soaking and defrosting. The masonwork that was half a brick in depth was destroyed in some parts of the wall. The upper layers of the decorative towers, the masonwork and side frontons, and even separate bricks of the masonwork from the outside walls were also destroyed by soaking and defrosting.

Deep cracks cover the outside walls of the wing: the crack opening is 5–40 mm along the whole height of the building. The prospecting, next to the wing wall near the moat, was worked out to find the reasons for the outside walls' cracking. Brickwork of the castle wall was found during the pit work (the depth was 0.3–0.35 m) between the entrance and the north towers. The positioning of the foundations of the castle's wall proves that a part of the outside wall of the wing was sited on them. The stone wing was built on the site of the former building. So we can suppose that the wing's

FIGURE 11.11 Wing and its outbuilding (from the castle yard). (Photo taken by the authors.)

foundation is made up of several parts, which were built in different times. An uneven foundation settlement is a probable reason for the appearance of cracks. The technical state of the walls is found to be unsatisfactory.

An examination of the wooden walls and the garret floor beams showed axial cracking (filament tearing) of the beams. A compression and bend test of the samples was carried out to find the strength properties of the wooden walls and garret floor beams. Wooden samples were sawn out of beams and logs. The compressive strength along the fibres of the samples was 45–49 MPa and the bending strength was 35–39 MPa. The technical state of the log walls and the garret floor beams is satisfactory. The compressive property of the wood is enough to bear the actual and assumed loads.

Truss systems with lathing, roof covering made out of asbestos and cement sheets of the wing and the outbuildings are considered of poor quality. There are cracks, spalls and gaps in the joints of the roofing sheets. Many of the rafters and lathing are rotten because of soaking. There is no gully hole, drip cap or blind areas. The results of sensors data analysis showed no development in the crack opening in the brickwork of the outside walls. However, the destruction of the outer blockwork and the roof frame are progressively dangerous.

The following means are required for the safety and further use of the wing and the outbuilding: restoration of the destroyed parts of the blockwork and socle; change the roof frame roofing; install metal drip caps; fill the cracks with cement-sand mortar brand of nothing less than M100; make a drain system from the roof and the blind area.

11.3.3 3D Model of Architecture and Size of Castle Complex

Three-dimensional modelling of the Lubcha Castle complex was made in a '3Ds MAX' programme using the data of photo documents, drawings of the entrance and the south towers, the bed of the north tower, the wing with the outbuilding and the palace itself. A 3D model is required to have a complete view of the whole architectural complex when placed side by side with the restored buildings of different eras (Bakatovich et al., 2011).

In accordance with the 3D model, it is planned to restore the north tower on the preserved foundation and the sections of the wall between the north and the entry towers (Figure 11.12). Restoration of the wall parts between the towers will make it possible to preserve the composition of the complex and will be used as observation areas.

Restoration of the fourth eastern tower and the castle wall up to the south tower will complete the complex. Firstly, the restoration of a rubble tower foundation is required. The log frame of the tower is planned to be three tiers with loopholes in the two lower towers and an observation area on the top tier. The roof cover will be shingled. The castle's defence wall between the east and south towers will be palisade with a cloister. It is planned to create another entrance gate in the restored walls and also to reconstruct the original wooden bridge from the main entrance of the tower over the moat rather than the original earth fill.

A walking area will be created along the northern facade of the palace using the technique of planking on wooden bars leaning on wooden pillars. The area will be decorated with a wooden decorative fence. It is also planned to build two stairs with observation areas leading to a reconstructed wooden pier.

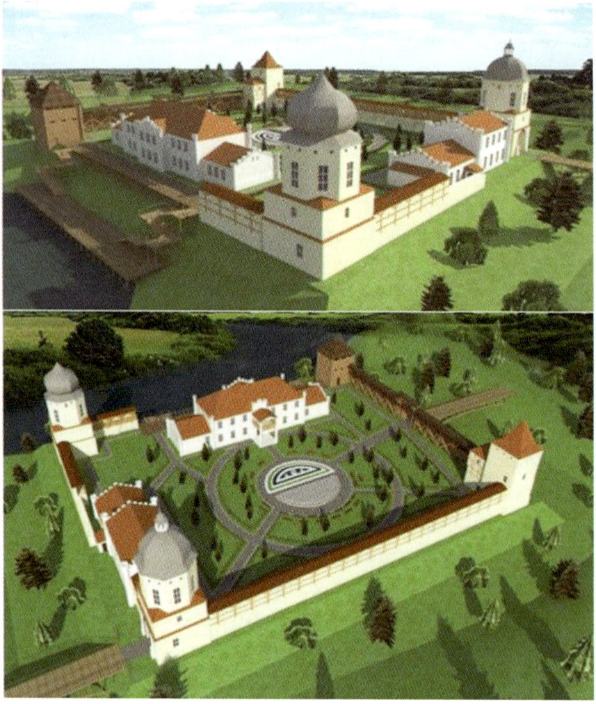

FIGURE 11.12 3D model of Lubcha Castle complex. (Photo taken by the authors.)

Improvements to the yard are shown in the plan (Figure 11.12). The carriage way is planned to imitate a closed ring of 4.5 m wide. The interior of the small circle is divided into two areas by a pathway. The first recreation area is a partially reconstructed area of the former castle territory and has semicircle fountains with small architectural elements. The other area, a semicircle covered with rubble, is a theatre space with a semicircle wooden or stone stage, opposite the fountain across the pathway. The amplitude of the ground makes it possible to organise historical and theatrical performances and cultural events, thereby attracting tourists to Lubcha Castle.

The 3D model of an architectural and spatial composition became the basis for the castle complex restoration project and is now funded by a charity fund 'Lubcha Castle' that is well known for supporting restoration work.

11.4 CONCLUSION

1. The use of nondestructive test methods during the all-round technical examination of the construction made it possible to outline a proper set of solutions for the reparation and restoration work and also provide a reliable service to Peter the Great's house.

2. The premises were improved after the restoration placed a new decorated fence all around the territory. Since 2014, there has been a permanent exhibition in the house that once belonged to Peter the Great.
3. Pupils who are interested in the city's history can attend workshops that take place in the basement. These pupils still have the opportunity to study precious museum items that describe the culture and lifestyle of the Polotsk citizens in the nineteenth and the beginning of the twentieth century.
4. The entrance and south towers, as well as a defence wall between them, have been completely restored in Lubcha Castle in recent years. The recommendations of the specialists who carried out a technical examination of the defences by means of nondestructive test methods, were taken into account especially when considering the restoration works of the entrance and south towers.
5. In 2016, the base of the north tower was found (using georadar), which simplified the restoration work of the bastion. This shows the importance of using the georadar sensing method in a comprehensive archaeological examination of the castle and its territory.
6. The results of a technical examination showed considerable damage to the wing and its outbuilding. It is important to follow the recommendations in Section 11.3.2.2 to preserve the construction. The current technical state of the castle construction can be used to improve the original exterior and the architectural peculiarities of the castle complex.
7. A 3D model of the castle complex is kept at the offices of the fund Lubcha Castle. The plans for the restoration work are based on the model or plan previously outlined. Spatial visualisation of the castle complex made it possible to evaluate the relevance of the architectural and construction decisions that had been made in the preliminary stages of planning.

REFERENCES

Bakatovich A., Davidenko N., Levadny A. and Lishtvan P. 2011. The concept of the Lyubcha castle development as a tourist center. *Herald of Polotsk State University. Series F. Civil Engineering. Applied Sciences. No.* 8., pp. 7–11.

Bakatovich A., Davidenko N., Lifshic U. and Pechinski I. 2008. Past and present of architectural monument of XVI ct.: Lubch castle complex above the Neman. *Herald of Polotsk State University. Series F. Civil Engineering Applied Sciences. No.* 12., pp. 4–12.

Bakatovich A., Ivanenko A., Shabanov D. 2007. Study of Peter the Great's house bearing and frame structures. Technical Report on Theme No. 27584. Polotsk State University, p. 74.

Barkun L. 1983. Historical landmarks of XVI–XIXctct. Castle in Lubcha, Novogrudok district Grodno region. Integrated Scientific Studies. Scientific-Restoration Studios. Ministry of Culture of BSSR. p. 40.

Chernavski I. and Saitgarzeva S. 1984. Lubcha castle. *Historical and Cultural Monuments of Belarus. No.2.*, pp. 46–50.

Kremnev A., Polakov A. and Pivovarova S. 2008. Georadar sensing of soil in construction engineering. *Collected Papers of XV International Scientific and Practical Seminar, Polotsk State University*, pp. 191–196.
Sinavski V. and Usoskai L. 2008. Restoration with reconstruction of historical and cultural heritage of 2nd category (Peter the Great's house) in Polotsk, Nizhne-Pokrovskaya Street, No. 33. Scientific and technical report. *Integrated Scientific Studies, UE Project Restoration*, p. 11.
Tkachov M. 1978. Defenses of western lands of Belarus XIII–XVIII ctct., Institute of Culture, Academy of Sciences of BSSR. Science and technology, p. 144.

12 Optical and Electromagnetic Sensing for the Inspection and Characterisation of Ancient Masonry Arch Bridges

Mercedes Solla, Belén Riveiro, Susana Lagüela and Iván Puente

CONTENTS

12.1 INTRODUCTION

Masonry arch bridges have been used in engineering for thousands of years and still remain a vital part of the highway systems and railway networks throughout Europe. As an example, in the United Kingdom alone, there are of the order of 70,000 masonry arch spans (Ashour and Garrity 1998); meanwhile, more than 250 units can be found in north-western Spain (Alvarado et al. 1989), impacting on its rural landscape and representing a priceless cultural heritage for mankind. However, these aged structures are facing a number of challenges associated with potentially destructive conditions and it is often asked whether they are still safe for carrying high volumes of modern traffic.

To provide accurate information to engineers for the inspection and characterisation of these old bridges, to carefully maintain them in the future and to analyse their structural behaviour in a more realistic manner, it may be necessary to develop up-to-date methods, rather than depending on old design methods or old codes of practice. In this regard, nondestructive testing (NDT) has recently increased in importance in the detection of defects and anomalies in historical buildings, being additionally supported by the International Council on Monuments and Sites (ICOMOS) and the United Nations Educational, Scientific and Cultural Organisation (UNESCO) (Puente et al. 2015). Nevertheless, each NDT method provides different information on the masonry structures; therefore, this chapter presents an overview of the assessment of masonry arch bridges, combining three NDT techniques: ground-penetrating radar (GPR) and infrared thermography (IRT) are applied in the detection of pathologies in the inner layers of the bridge, thus providing valuable information about hidden geometry, moisture content and filling conditions. On the other hand, terrestrial laser scanning (TLS) can quickly measure the visible geometry of the bridge, including its superficial pathologies, giving extra information about their sizes and depths and the latest indicators of their severity.

This chapter presents an overview of different NDT methods, as well as the data acquisition strategies for each method and the subsequent processing techniques. Then, some examples of the results obtained by the application of each of these nondestructive methods, either individually or in combination, to masonry bridges are reported.

12.2 METHODS AND METHODOLOGIES FOR DATA ACQUISITION AND PROCESSING

12.2.1 Terrestrial Laser Scanning

Laser scanning is a geomatic method that allows the obtaining of the 3D geometry of objects' surface in an automatic way by using light detection and ranging (LiDAR) technology. Different classifications are standardised for laser scanning, attending to the different principles used for distance measurement (Vosselman and Maas 2010). For civil engineering applications, the most common method is based on measuring the time of flight (TOF) of the laser beam. This principle allows distances to be remotely computed by measuring the time delay between the emission and the return of a laser beam that travels from the instrument (a laser scanner) to the object. This distance may be obtained directly by using short repetitive laser pulses, or indirectly by modulating the power of the wavelength of the laser beam and using phase difference (Vosselman and Maas 2010).

To efficiently cover the whole field of view of the instrument, laser scanners contain mirrors to deflect the laser beam in different directions. TLS (static scanners) allow directing the laser in vertical directions (rotation around the horizontal axis), while rotation around the vertical axis may be done through another mirror or through a mechanism. This mirror can rotate at a very high speed through the whole vertical field of view (360°). As a result, a very dense point cloud is obtained. This point cloud basically contains the *X*, *Y* and *Z* coordinates of each surface point, and some other attributes such as the amplitude of the laser beam or colour data, when red-green-blue (RGB) cameras are attached to the scanner.

The data processing normally depends on subsequent applications of the point clouds collected from the object. A point cloud can be defined as the most primitive 3D model that is directly recorded by the scanner. This model is composed of discrete and isolated points; however, the reconstruction of 3D models usually refers to building continuous surface models.

A digital surface model (DSM) is normally computed by triangulation or another interpolation method. A common procedure to create the mesh is based on triangulation.

The work related to documenting a historic construction sometimes requires colour information to assist in the process of documentation. This necessity entails a previous process of digital image registration. If a photorealistic 3D surface model is obtained, orthoimages are produced by means of a differential rectification procedure. The differential rectification consists of the individual projective transformation of each triangle of the DSM on a reference plane. After the geometric transformation, densities require to be assigned through a radiometric interpolation. The resultant orthoimages can be exported to computer-aided design (CAD) programmes to be used as the basis for the documentation plans of the structures.

Sometimes, with the aim of a structural assessment of an historical construction, a detailed 3D model with all the relevant features regarding complex structural geometry is required. Such a detailed geometric model could then be directly used as the basis for the development of a more complex numerical model, for example, using the finite element method (FEM). In those cases, surface-based representations need to be converted into volumetric representations. However, the triangular meshes previously described might sometimes be inefficient for the aims of structural engineering applications. Their principal limitations come from the presence of gaps, excessive sharp angles, and so on, that may cause problems during the mesh generation, thereby preventing convergence solutions.

Until now, three main approaches have been developed to allow for the creation of a detailed 3D CAD model from an initial point cloud.

The first usual approach consists of the re-edition of the mesh model into an intermediate CAD software before being imported into the FEM packages. CAD software prevents mesh-generation problems because these programmes rely on straight lines and generate closed models (Laefer et al. 2011). The modelling process begins with the extraction of feature lines that describe the main characteristics of the model through cross sections. Such cross sections are then used to perform extrusions and obtain a final CAD model. As reported by Tang et al. (2010), this approach has the advantage of not demanding excessive computational resources, but it can lead to errors when the components do not follow their idealised geometries.

A simpler approach consists of creating the 3D solid model directly from the raw point data, avoiding possible errors caused by poor smoothing or sharp edges during triangular mesh generation. The focus is then to work with the global registered point cloud, importing it into an intermediate CAD software for further 3D modelling. Once the point cloud is imported into the CAD platform, the process of modelling is straightforward using primitive fitting tools, cross-sectional methods and extrusion techniques. Usually, such operations are based on the constraints imposed by the point cloud. The CAD software must incorporate specific modules for point cloud visualisation when dense point clouds need to be handled. Otherwise, it would be desirable to extract orthogonal views from the point cloud and just draw above the points of such a 'slice' to create the desired part of the CAD model. By repeating this and assembling all parts, the entire 3D model could also be achieved.

The third and final method is based on the creation of nonuniform rational B-splines (NURBS) upon a given triangular mesh.

The first two methods have as their main drawback the time required to perform the process, which is subject to manual operation and requires highly specialized skills in geomatics, even though there is subtle idealisation of the model. The NURBS method has the main advantages of being particularly precise for complex geometries, its exportation to FEM packages does not present the aforementioned problems and it requires less human work; however, it presents a high computational cost. To sum up, depending on the accuracy sought, the available time and computer performance, the user is responsible for choosing any of the three methods or even mixing them when composing the 3D model.

12.2.2 Infrared Thermography

IRT is a physical method based on the capacity to measure temperature values from the infrared radiation emitted by bodies. When bodies are at temperatures above 0 K, their constituent atoms vibrate. This vibration implies the emission of radiation with a wavelength of between 7 and 14 μm, which belongs to the band of the spectrum named *thermal infrared* (TIR). The higher the temperature of the body, the faster the atom vibrates, and consequently, the greater the quantity of infrared radiation emitted.

Thermographic cameras consist of specific lenses, mostly made of amorphous silicium or vanadium, which are only sensitive to radiation in the TIR band of the spectrum. The quantity of radiation is measured as an electric signal, and then processed in the thermographic camera in such a way that the user receives a temperature value per image pixel. A colour palette can be applied to the existing temperatures to make their understanding and interpretation easier (Figure 12.1). More information about the principles of the technique can be found in Meola and Carlomagno (2004).

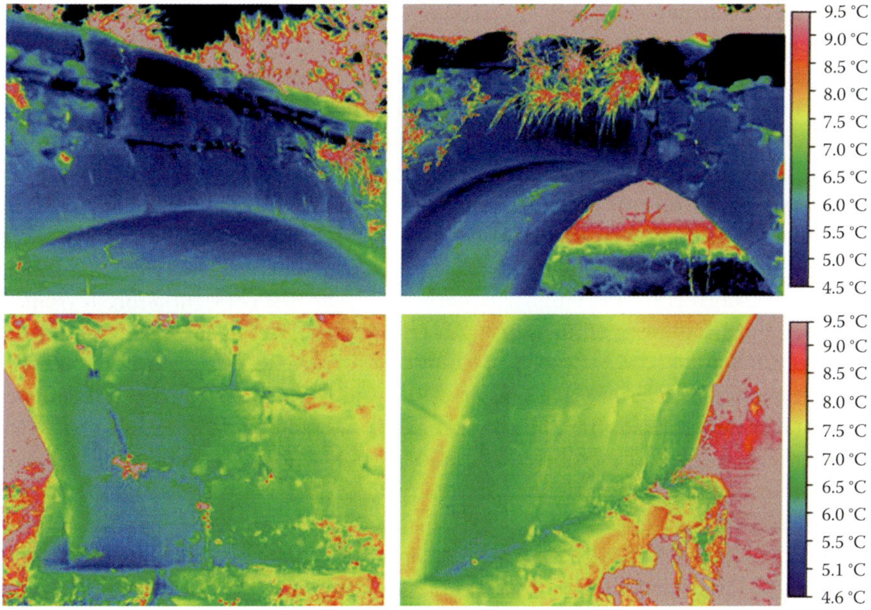

FIGURE 12.1 Example of thermographic images acquired from a masonry bridge, with their colour palette.

There are different recommendations when performing a thermographic inspection, according to the type of element under study and the objective of the inspection. This way, if the focus is set on the detection of thermal anomalies without further measurement, the technique is called *qualitative thermography*, while if knowledge of the real temperatures is needed for processing and geometric correlations, the technique is called *quantitative thermography*. On the other hand, if there is an artificial thermal excitation of the body under study, the work is called *active thermography*, whereas the acquisition of thermal information from the state of the body 'as is', with no incidence of exterior thermal contribution, is defined as *passive thermography* (Lagüela et al. 2012a).

Regarding bridge inspections, their size and outdoor location complicate the application of the active thermographic technique, due to the difficulty of finding a versatile heating source to apply in outdoor scenes. However, the detection of most thermal pathologies benefits from thermal excitation to highlight their presence, given that these types of anomalies present a different thermal behaviour (thermal conductivity and diffusivity) with respect to the nondefective areas. For this reason, the common approach is to use a solar incidence as the heating source, as in Lagüela et al. (2012a).

Related to this, meteorological conditions have to be taken into account when performing a thermographic inspection, especially in outdoor scenes. Thus, a direct solar incidence should be avoided, in order to eliminate the measurement of differential heating during the inspection. The effect of a direct solar incidence can cause existing thermal pathologies of small dimensions or low superficial incidence to be overlooked (Martín et al. 2004). Given that the effect of the sun is useful for detecting thermal pathologies, but not with a direct incidence, the optimal time range to perform these thermographic studies is after sunset, or at least after the sun reaches the surface of the bridge. Furthermore, the presence of water on the surface affects the thermographic measurement because it homogenises the temperatures. For this reason, thermographic inspections should be performed after a minimum of 24 h has elapsed since the last rain period (Rosina and Spodek 2003). In addition to the meteorological considerations, a valid thermographic inspection requires the introduction of two corrections in order to have a more accurate temperature measurement: atmospheric correction and emissivity correction. The first correction implies taking into account the ambient temperature and relative humidity in order to calculate the effect of the atmosphere between the camera and the object under study. This effect is dual: on the one hand, an attenuation of the radiation emitted by the object under study; on the other hand, the introduction of the radiation from the atmosphere into the measurement. Most thermographic cameras perform the atmospheric correction directly, through the introduction of the temperature and relative temperature values, as well as the distance between the camera and object (Hamrelius 1992). It results in a coefficient from 0 to 1, with a common value of 0.95–0.99 for terrestrial acquisition and normal atmospheric conditions. With increasing temperature and humidity, the correction coefficient decreases, although some authors consider it constant for each thermal scene (Yang and Li 2009).

The emissivity value of each surface has an important effect on the thermographic measurement, in such a way that the apparent temperature is very different from the real temperature (Lewis et al. 1973). This parameter is a measure of the radiation emitted by the real body with respect to the radiation emitted if the object was a black body. Its value mainly depends on the superficial finish, its level of rusting and brightness, and other conditions causing degradation (Luo et al. 2014). This way, the emissivity value of a black body is 1, while real bodies have emissivities from 0.1 to 0.99. In general, construction materials present emissivity values between 0.90 and 0.96, in such a way that their inspection by measuring apparent temperatures does not imply a great error with respect to the real temperature. However, it is possible to perform the emissivity correction after the inspection, by applying the correction to the temperature values measured, according to Stefan Boltzmann's law (Rodríguez-Martín et al. 2016).

12.2.3 Ground-Penetrating Radar

GPR is a geophysical method based on the propagation of a very short electromagnetic pulse (1–20 ns) in the frequency band of 10 MHz–2.5 GHz. The GPR antenna is moved over the investigation target in a specific direction with a constant distance between the transmitter and receiver. The transmitting antenna emits an electromagnetic pulse into the ground, which is partly reflected when it encounters media with different dielectric properties and is partly transmitted into deeper layers. Next, the reflected signal or trace (a radar wave containing all the reflections in depth) is returned to the receiving antenna. Nevertheless, the control unit enables the processing of all the individual traces (A-Scans) into a 2D image (B-Scan or radargram) by applying a colour palette to the amplitude (strength or intensity of the reflection) values registered. Commonly, a greyscale colour palette is applied, which assigns the black colour to the minimum amplitude and the white colour to the maximum amplitude. The intermediate values are converted into grey shades of different tonality. Thus, by moving the antenna over the ground, an image of the shallow subsurface under the displacement line is obtained. These 2D images, called *radargrams*, are *XZ* graphic representations of the reflections detected. The *X*-axis represents the antenna displacement along the survey line, and the *Z*-axis represents the two-way travel time of the pulse emitted (in terms of nanoseconds). If the time required for the electromagnetic pulse to go from the transmitting antenna to the reflector into the ground and return to the receiving antenna is measured and the velocity of this pulse in the subsurface medium is known, then the position of the reflector can be determined. More details on the basic principles of GPR can be found in Annan (2003) and Daniels (2004).

The GPR system used in this work was a RAMAC/GPR system from MALA GEOSCIENCE. An important parameter controlling the depth range of GPR is the transmitting antenna frequency. The antenna frequency employed in a GPR survey should be carefully chosen because a balance must be maintained between a low frequency antenna, which gives deeper signal penetration but poorer resolution, and a higher frequency antenna, which gives better resolution but shallower penetration. The 250 and 500 MHz central frequencies were selected as the most suitable to evaluate masonry arch bridges due to their optimal compromise between penetration and resolution. The purpose was to reach the bridge foundations while analysing the shallower filling.

The GPR data acquisition was based on constant distance intervals using the common offset mode with the antenna polarisation perpendicular to the data collection direction.

In most cases, for the 250 MHz data, the distance interval (or in-line spacing between traces) and the total time window were equal to 5 cm and 200 ns defined by 516 samples per trace, respectively. For the 500 MHz data, the distance interval was equal to 2 cm, and the total time window was set to 100 ns, which was also defined by 678 samples per trace.

To measure the profile lengths and to control the distance interval, an odometer wheel was attached to the antenna. However, some of these masonry arch bridges had intact ancient flagstone surfaces composed of large irregular blocks, which made measurements difficult as the continuous movement of the survey wheel was interrupted. In such circumstances, the wheel should be calibrated on-site to avoid inaccuracies in spatial trace positioning.

Masonry arch bridges are complex and heterogeneous structures that have undergone several restorations using different materials. Therefore, it is recommended to collect more than one longitudinal profile to the main direction of the bridge. Moreover, the GPR profiles should be collected an appropriate distance from the parapets of the bridge in order to avoid airwave events associated with the reflected signal.

There are several GPR processing filters to reduce clutter and any unwanted noise. The objective is to correct the down-shifting of the radar section due to the air–ground interface and to amplify the received signal, as well as to remove both low and high frequency noise in the vertical and horizontal directions. For signal processing, the ReflexW software (Sandmeier 2016) was used in this work and the filters most commonly employed were as follows.

Time-zero correction. Traces require adjusting to a common time-zero position since thermal drift, electronic instability and cable length differences or variations in an antenna air gap can cause 'jumps' in the air/ground wavelength first arrival time (Olhoeft 2000).

Vertical or temporal filtering (*subtract direct current [dc]-shift*). A common feature in commercialised GPR equipment is the appearance of a continuous or very low frequency component (dc component) in the traces recorded by the radar, since the averaged level of the signal is moved from zero amplitude to a different value. The appearance of this component is usually associated with either inductive phenomena or possible instrumentation dynamic range limitations (Annan 2003).

Gain. During the propagation of the radar signal through the subsurface media, several factors result in a decrease in the signal strength and the electromagnetic signal will experience attenuation (ground materials, frequency antenna, etc.). Gaining consists of amplifying the received signal by multiplying the data using a mathematical function or manually entering gain values.

Horizontal or spatial filtering (*subtracting average*). The main objective of these filters is to remove potential low frequency noise which appears in the form of continuous horizontal bands along the recorded traces or only in some parts. Spatial filters are good at removing noise originating from bad coupling between an antenna and a medium and to eliminate ringing from the data sets.

Bandpass filtering (*Butterworth*). Sometimes, the energy of reflected signals is placed in a particular frequency range whereas most noise sources and background noise are in a different frequency range. Therefore, distinguishing noise from useful information is often possible in the frequency domain. A bandpass filter is a combination of high-pass filters – for removing signal drift and low frequency noise – and low-pass filters – for high frequency noise.

Topographic corrections. Historical bridges usually present a medieval double slope or a steep profile. Topographic corrections are applied when the bridge has an upward arching profile to improve the accuracy of imaging subsurface features. Without this correction, the location of the subsurface features can change and the shape of the features becomes distorted.

12.3 RESULTS PRODUCED BY EACH METHOD

12.3.1 Terrestrial Laser Scanning

When using TLS for documenting structures, the main result is a geometric model. Depending on the particular application, the model can be given in the form of a 2D or a 3D model that is composed of isolated points (Figure 12.2a), connected lines or volumetric shapes (Figure 12.2b). The following sections show some common applications related to the documentation of masonry arch bridges; however, the spectrum of applications can be wider.

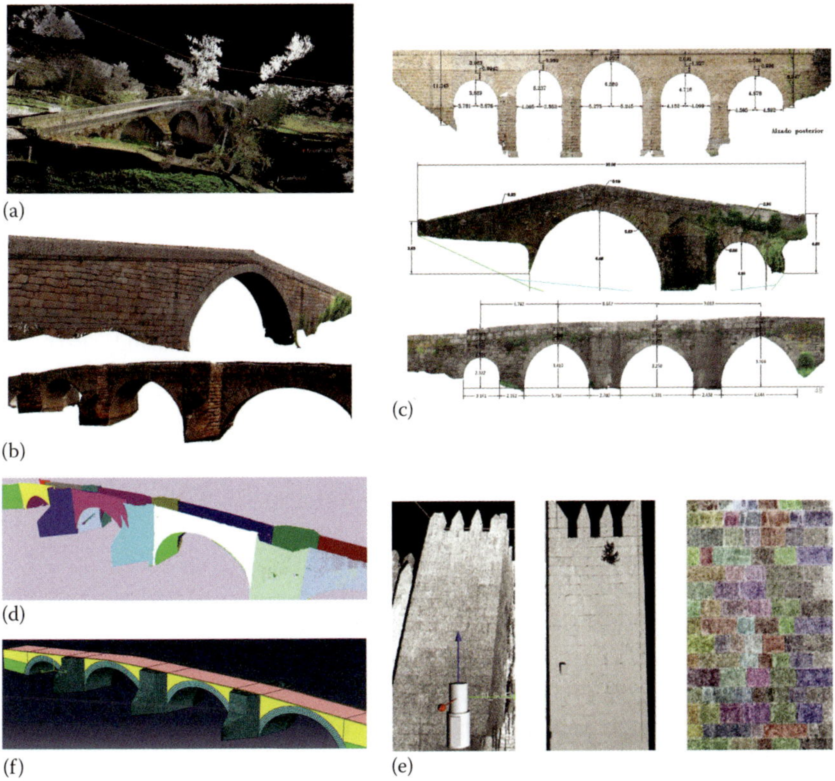

FIGURE 12.2 Results of TLS survey: (a) raw point cloud (coloured) of a masonry arch bridge and its surroundings; (b) 3D surface models of masonry arch bridges; (c) metric analysis over orthophotos created from 3D textured surface models; (d) results of the automated segmentation of a bridge point cloud; (e) automatic segmentation of masonry units from a point cloud; (f) FEM model created from the combination of a 3D volumetric model created from TLS and the material characterisation using GPR.

12.3.1.1 Dimensional Analysis of Structural Members

TLS represents great potential to perform geometric analysis thanks to the amount of detailed data recorded. Using the orthoimages derived from photorealistic models, it is possible to perform dimensional analysis using CAD software (Figure 12.2c), as well as accurately detect pathologies or signs related to the stability of the constructions. This is the case of a survey performed in the Roman Bridge of Segura (Spain) where a dimensional analysis procedure was performed and some of the most important geometric-constructive parameters could be estimated. What is more, pathological structural analysis was completed with the extraction of sections in the three main directions of the bridge. The most relevant damage detected was asymmetries in one of the arches, a skew in one of the bridge's bowls (one of those supporting the asymmetrical arch) and significant deformation in one of the tympanum walls.

12.3.1.2 Geometrical Modelling for Structural Analysis Purposes

Historic structures normally have singular geometries that are intrinsically related to their structural behaviour. Therefore, it is compulsory to obtain the geometry of the construction as it actually is because this geometric model will directly contribute to the precision of the structural model used for the structural evaluation. Historic arch bridges are a clear example where geometry represents an important issue, but in general, this approach is extendable to any historic masonry building. Among the different approaches used to create geometric models for the structural analysis of masonry arch bridges, in the work presented by Stavroulaki et al. (2016) a 3D model built from laser scanning of a masonry arch bridge was used as the basis for a structural modelling using the finite element method. On the other hand, also related to the evaluation of bridges, Riveiro et al. (2011, 2013) have demonstrated the suitability of accurate 3D models of arch bridges to perform the subsequent structural analysis in the framework of the limit analysis theory. Thus, after creating the DSM of arches (extracting cross sections of the 3D model or directly working with orthophotos), it was possible to define the geometry required to conduct a stability analysis of the arches using the rigid blocks method.

12.3.1.3 Automation in the Processing of LiDAR Data for Structural Applications

A growing interest in the automatic creation of structural models from laser scanning data is increasingly reported in the literature (Truong-Hong et al. 2013, Cabaleiro et al. 2014, Barazzetti et al. 2015). Regarding masonry bridges, Riveiro et al. (2015, 2016) presented works where the point clouds collected by laser scanning systems can be directly interpreted without the intervention of human operators. The final goal of these methodologies is to avoid the specialised training required for handling geomatic data, so that the structural engineers can directly use the data for the structural modelling. In this sense, the work of Riveiro et al. (2016) presents a methodology for the automated segmentation (partition) of large point clouds of masonry arch bridges into its main structural elements (Figure 12.2d), namely: spandrel walls, arches, piers, cutwaters, pathway, etc. This methodology was validated in ten Spanish and British arch bridges. Another methodology presented by Riveiro et al. (2015) refers to the segmentation of point clouds of masonry walls into masonry units and joints, whose final aim was to have the necessary information to automatically create structural models that best represent the assembly of stones having accurate information on the geometry of the masonry blocks and the characteristics of interfaces, as shown in Figure 12.2e.

12.3.2 Infrared Thermography

IRT has been tested as an inspection technique in different masonry structures, among which the Lubiáns Bridge can be found (Lagüela et al. 2012b). The versatility of the technique allows its application in two types of defects: superficial and sub-superficial. As explained in Section 12.2.2, the detection of each type depends on the conditions of the acquisition, such as the presence of thermal excitation, time window of the inspection and ambient conditions.

The superficial nature of the technique allows the collection of information about the external state of the bridge, highlighting some pathologies that cannot be distinguished through a visual inspection, such as cracks and the presence of vegetation and moss. The detection of these pathologies is based on the different temperatures of vegetation and moss (and their roots) as living beings, in comparison with masonry. Thus, the automatic detection of these elements is straightforward because their temperature is always around 30°C–35°C. However, some attention should be given to the ambient conditions, since the masonry can achieve the same or higher temperature values on warm, sunny days.

In addition, the different thermal response of the materials allows the detection of pathologies in the subsurface, mainly consisting of humidity and air cameras present under the superficial material. The reason for this detection is the different thermal conductivity and diffusivity of water and air compared with masonry, which imply a different thermal response to thermal stimulation. Given that the detection of subsurface defects requires an external thermal excitation, solar radiation is usually the heat source used for this task in bridges.

12.3.2.1 Thermographic Inspection

The application of the thermographic technique, by itself, is not complicated and does not need specific post-processing. In fact, the most important aspect is data acquisition, because its performance in not-optimal ambient conditions can provoke the failure to detect pathologies because they are not sufficiently highlighted, or even because they are hidden under other effects such as water covering the surface and homogenisation of the superficial temperature. After data acquisition, emissivity correction should be performed if the absolute temperature values are of interest for the study. Usually, ambient correction is introduced in the camera so that images acquired already include this variation in the measured temperature values.

Thus, when all corrections needed are applied, data interpretation and detection and the identification of pathologies is the most complicated task, which requires the performance of an expert in both heat transfer and the materials and structures under study. Recent trends have focused on the automation of the interpretation process through the introduction of thermal constraints in the post-processing. As an example, Cereijo et al. (2014) developed an algorithm for the automatic detection of thermal bridges in buildings; although their detection of the pathology reached 80% accuracy for most cases, it also presented 50% false positives, resulting in the overdetection of thermal bridges.

12.3.2.2 Thermographic and Geometric Inspection

The conjoint analysis of thermographic images with geometric or GPR data requires further data processing to make thermal data adequate for its combination. In these cases, first, a geometric calibration of the camera should be performed to allow the correction of the distortion introduced in the images by the lens. With this aim, different calibration grids have been designed, based on differences in temperature (Lagüela et al. 2011) and differences in emissivity or apparent temperature (Lagüela et al. 2012).

Once images are geometrically corrected, there are two main options for analysis: one is the generation of a thermographic mosaic in order to have one image covering each surface under study. This product allows a more accurate detection of pathologies that could be missed on an image-by-image study. This mosaic is generated through the automatic detection of image features or points of interest, their matching and coregistration. When the overlapping area between images is sufficiently large, it is possible to generate a 3D point cloud directly from the thermographic images (Lagüela et al. 2012c). However, this task is complicated due to the low resolution of these images (640 × 480 is the largest sensor in the market for uncooled thermographic cameras).

The second option is only possible when geometric data from the bridge is available in point cloud format. Thus, the approach consists of the registration, image by image, of the thermographic information with the point cloud. This generates a thermographic point cloud, which improves the understanding of the phenomena occurring in the bridge through a 3D perspective. The registration can be manual or automatic (Lagüela et al. 2013), making the second option the most efficient. Regarding automation, features for matching can be points or lines, making the first more commonly known while the second provides more consistent results.

The interpretation of the images, together with the direct visualisation of the structure under study, improves the understanding of the phenomena occurring, and consequently the behaviour of the bridge, its pathologies and the presence of fluids. In Lagüela et al. (2012b), it was possible to understand the preference paths of water in Lubiáns bridge thanks to the availability of thermographic and geometric information.

12.3.3 Ground-Penetrating Radar

The GPR technique has demonstrated the ability to provide information about internal bridge details. Some examples of the application and information obtained on masonry arch bridges in Galicia (northwest of Spain) are briefly explained.

12.3.3.1 Documentation from a Historical Point of View: Former Geometries, Restorations and Reconstructions

As relevant results, it was possible to identify interesting information such as the presence of possible hidden arches in the case of San Antón Bridge (Figure 12.3a) or a different historic shape to the structure as in the Roman Lugo Bridge (Figure 12.3b). In the case of the San Antón Bridge, three large upward arching reflections were visible. However, the bridge has only two visible arches and this unexpected third reflection is probably generated from an arch now hidden from view by later construction. Below this unexpected reflection there is an additional reflection most probably associated with the river bed of a different river channel in former times. Observing the polarity of the arch reflections, it is important to mention the inversion on the polarity of this third reflection regarding the other two visible arches. Accordingly, this third arch is not empty and is probably filled with some material of a lower density than the granitic masonry of the arch. Moreover, the filling did not produce

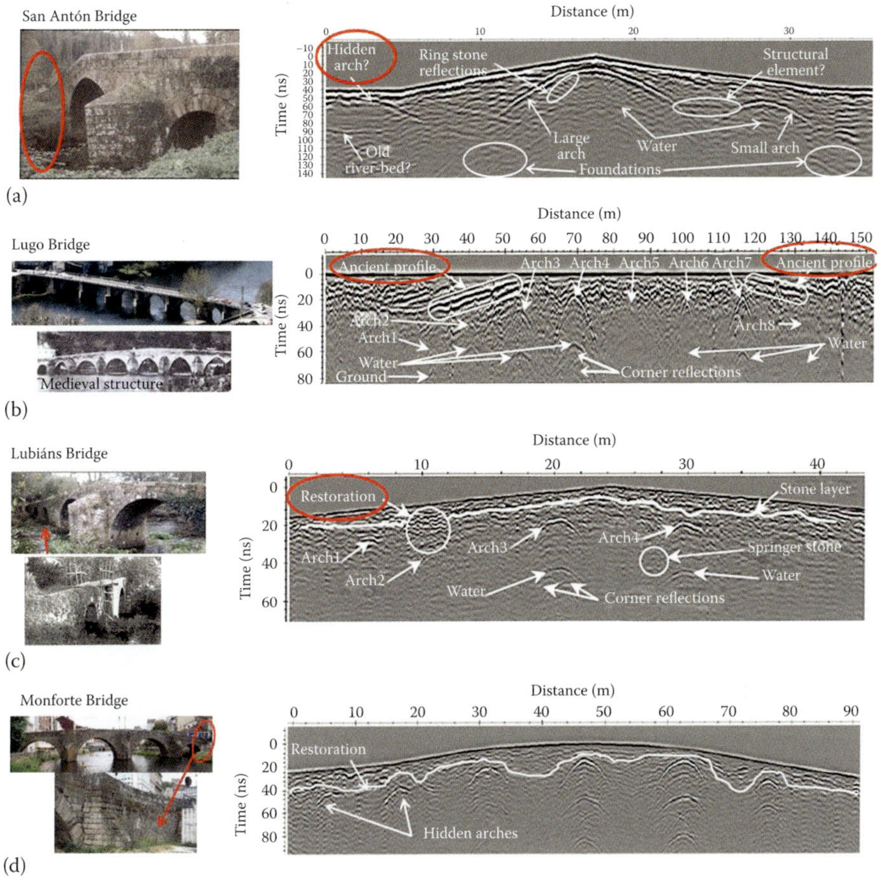

FIGURE 12.3 GPR examples of application in masonry arch bridges showing valuable information from a historical point of view. (a) Existence of a hidden arch, (b) existence of ancient profiles of the structure, (c) detection of an arch reconstruction and (d) existence of hidden arches and refilled stonework.

reflections, which suggests it is quite homogeneous in composition. Regarding the ancient shapes, in some medieval bridges, the former double slope profile had been filled to create a horizontal pathway, and the fill materials used were different to the original ones, presenting sufficient dielectric contrast to be distinguished in the data. This is the case of the Lugo Bridge showing a slightly sloping constant reflector at both margins of the structure. Additionally, it was possible to detect and map the remains of restoration tasks throughout time, such as the reconstruction of arches (Lubiáns Bridge in Figure 12.3c). In the case of the restored arch, a different reflection pattern is observed just above the arch. Historical references inform that this vault was rebuilt from ruins in 1984, breaking the continuity of the pavement with the reconstruction.

Finally, Figure 12.3d presents the case study of Monforte Bridge. It was interpreted that this bridge could be emptied and later refilled using different backfill than the original one. Moreover, it was possible to detect six hyperbolic reflections produced by the arch–air interface of the arches. Although the visible bridge structure has only four arches, two more hyperbolic reflections were identified at the beginning of the radargram, which may correspond to the two nonvisible arches hidden by the construction of a breakwater at this margin of the original structure. All these bridges were widely analysed in Solla et al. (2010, 2011a, 2014, 2016a).

12.3.3.2 Documentation from a Civil Structural Point of View: Pathologies and Construction Details

Additional information can be provided about the nature of a bridge's foundations, such as in the case of Areas Bridge (Solla et al. 2012a), which lies on bedrock and the reflection at the interface showed a rather strong signal due to the larger dielectric contrast between the media. Even a probable cavity was detected in one pier of Traba Bridge (Figure 12.4a), showing an anomaly with a strong signal from the high dielectric contrast between the masonry and the water filling the cavity (Solla et al. 2011b).

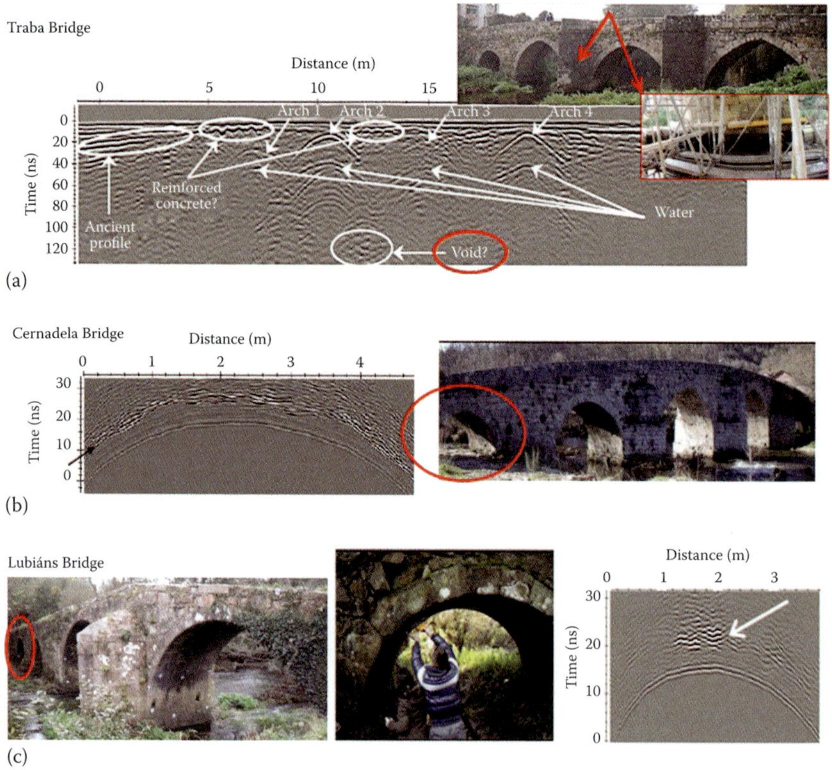

FIGURE 12.4 Examples of application in masonry arch bridges showing valuable information from a structural point of view. (a) Presence of a void in one pier, (b) ring-stone thickness and (c) detection of moisture problems.

Additionally, structural information was obtained such as the ring-stone thickness of arches as in the case of Cernadela Bridge (Figure 12.4b), where a hyperbolic reflection is produced by the fill–stone interface. Nevertheless, masonry arch bridges usually present nonhomogeneous filling due to the existence of different materials with different sizes, as well as the usual irregular geometry of the ring-stones in the extrados of the vault and probable ring separation. Therefore, while the identification of the interface between paving and filling is possible when collecting data over the pathway of the bridge, some difficulties can occur in detecting the ring-stone thickness due to the confluence of many unwanted reflections (Solla et al. 2012b). To avoid all these geometric irregularities at the fill–granite interface and clearly identify the ring-stone thickness, it is recommended to collect the GPR profiles through the intrados surface of the vault arch in the longitudinal direction to the structure. In such cases, because of the small size of the ring-stones, antennas with higher resolution are required. The 1 GHz antenna was commonly used with a distance interval equal to 1 cm, and a total time window of 30 ns, which was also defined by 343 samples per trace. Knowing the internal ring-stone thickness of the vault allowed for more accurate analytical calculations and a structural assessment of arches, such as in Solla et al. (2012b) and Stavroulaki et al. (2016).

Moisture problems in masonry can also be detected with the GPR method. As an example, Figure 12.4c shows the first arch from the left margin upstream of the Lubiáns Bridge, which has been identified as the most delicate because of the visible water exudations through the intrados of the vault. A GPR survey was also conducted through the intrados surface of the vault arch in the longitudinal direction to the structure to avoid scattering and to better understand how water is distributed through masonry and accumulated at the contact between the backfill and the arch-ring. Observing the radargram produced, a stronger reflection generated from the backfill–stone interface (white arrow in Figure 12.4c) was only clearly identified in the zones contiguous to the keystone of the arch (at $\approx$20 ns). The most probable explanation is that water could accumulate in this area.

The GPR data also allowed for the differentiation of building material in a structure, as in the case of the Roman Lugo Bridge (Solla et al. 2011a), in which vaults built with slate materials presented more attenuation than vaults built with granite. This provides interesting information from a structural point of view. Differences in stonework and filling can affect the structural reinforcement of a bridge in terms of its durability and strength.

12.4 COMBINED USE OF TECHNIQUES

Firstly, the integration of TLS and IRT data allowed for the construction of 3D models of the bridges with a thermographic texture. The primary 3D model produced from the TLS point cloud was textured with RGB images (Figure 12.5Aa). The next step consisted of registering the thermographic images to the previous 3D model for which the thermographic images have to be registered using the natural marks visible in both RGB and IRT images. Finally, IRT images were used to texture the 3D model of the bridge built from the dense point cloud previously obtained (Figure 12.5Ab). More explanation can be found in Lagüela et al. (2012b).

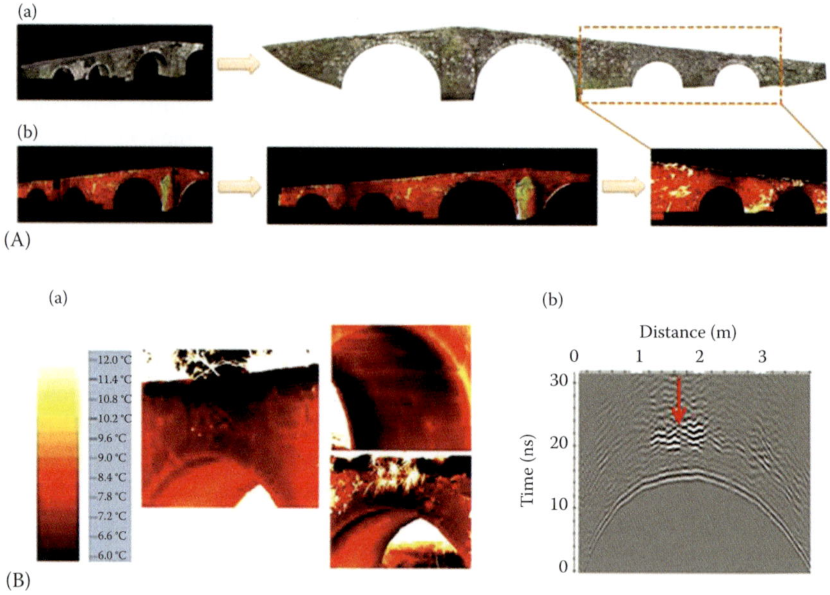

FIGURE 12.5 Results of the integral (A) TLS and IRT: 3D model with RGB (a) and thermographic (b) texture, and (B) IRT and GPR: Detection of moisture in masonry showing thermal images with a temperature interval from 6°C to 12°C (a), and 1 GHz GPR data produced (b).

Secondly, the TLS data can be used for the calibration of the velocity of propagation of the GPR waves in media. Knowledge of the most appropriate radar wave velocity allows for a more complete interpretation of the field data. Thus, the obtained value of the velocity of propagation in media is useful to estimate the inner building materials involved as well as probable moisture damages. As explained in Solla et al. (2014), velocity can be determined by adapting a diffraction hyperbola to the hyperbolic reflection produced by the arch–air interface. The accurate metric information of the arch span provided by TLS was used to define the hyperbolic shape to fit. A similar procedure, but from the difference pick application (Lubowiecka et al. 2011), can be used to characterise the filling when knowing the height of this filling over the arch.

Regarding the structural behaviour and stability of a bridge, integral TLS and GPR approaches will benefit the evaluation of these types of structures where arch thickness plays a key role in the ultimate load capacity of the structure. Different works in the published literature have demonstrated the potential of both techniques in the structural analysis of masonry arch bridges (Lubowiecka et al. 2011, Solla et al. 2016b, Stavroulaki et al. 2016). The TLS allowed for the production of 3D real models (external geometry) towards their implementation in more sophisticated analysis procedures involving numerical analysis such as finite element–based modelling. The GPR method provided the characterisation of the different materials in the stonework, and essential internal geometric parameters such as arch-ring thickness. According to this, it was possible to perform complex, but more realistic, stability analysis of a bridge.

Finally, for the evaluation of masonry arch bridges, the combination of IRT and GPR techniques can be useful to detect moisture in masonry. As an example, Figure 12.5B presents the results of both thermographic and GPR surveys over the first arch upstream of the Lubiáns Bridge, which has visible water exudations through the intrados surface of the vault.

The thermographic survey included the walls of the bridge (both upstream and downstream walls) and the inner part of the vault of the arch under study. The temperature interval selected was from 6°C to 12°C (colours ranging from dark red for the lowest temperatures to red, light red, orange, yellow and white for the highest temperature; Figure 12.5Ba). The thermal images facilitated the identification of probable moist areas in the bridge surface as those having lower temperatures (areas in dark red in Figure 12.5Ba). This is due to the cooling effect of evaporation on the surface where it takes place. Thus, important evidence of moisture appears on the upper part of the walls, near the border with the pathway and in the masonry over the arch.

To avoid scattering and complex reflection patterns produced by heterogeneous filling, a GPR survey was carried out passing a 1 GHz GPR antenna through the intrados surface of the vault in the longitudinal direction of the bridge. The GPR data provided information on the inner materials of the bridge. Observing the radargram produced, a stronger reflection generated from the backfill–stone interface (red arrow in Figure 12.5Bb) was only clearly identified in the zones contiguous to the keystone of the arch (at ≈20 ns). The most probable explanation is that water could accumulate in this area. This GPR interpretation agreed with the thermographic data obtained, which showed a more critical presence of water (dark red areas) at the surroundings of the keystone.

12.5 CONCLUSIONS

TLS is a very appropriate technology to reconstruct the overall 3D geometry of masonry arch bridges in an accurate and detailed manner; moreover, without requiring direct contact with the structure. From the 3D model created, 2D plans and orthoimages can be derived to perform further dimensional analysis. Some external pathologies and structural faults can also be assessed thanks to the accurate and detailed information registered, in combination with customised automated processing tools (i.e., moisture, lack of ashlar in masonry, detection of different construction materials, etc.).

For the analysis of an internal structure, this work proposes the use of IRT and GPR nondestructive techniques. IRT has demonstrated its capabilities to map moisture, like other near-surface (2–5 cm in depth) pathologies such as cracks and hollows in masonry. In addition, the GPR method has the ability to provide previously unknown information from the interior of bridges (from the surface to the foundations). Interesting aspects can be obtained from historical/archaeological and structural points of view (i.e., hidden arches, ancient geometries, cavities, differences in building materials, ring-stone thicknesses, etc.).

The combination of the three different methods has demonstrated its potential for a complete analysis of a bridge's structure from the exterior to the foundations. For example, the combination of TLS and GPR enabled more realistic and reliable analysis to be done while the combination of IRT and GPR allowed for the detection and mapping of moisture in masonry as well as its most likely origin and distribution.

REFERENCES

Alvarado, S., M. Durán, and C. Nárdiz. 1989. *Historical Arch Bridges in Galicia*. Official Association of Civil Engineers, Xunta de Galicia (in Spanish).

Annan, P. 2003. *Ground Penetrating Radar: Principles, Procedures and Applications*. Sensors & Software Inc. Mississauga, Canada.

Ashour, A.F. and S.W. Garrity. 1998. An upper bound analysis for the strength assessment of masonry arch bridges. Arch bridges, history, analysis, assessment, maintenance and repair. In *Proceedings of the Second International Arch Bridge Conference*, ed. A. Sinopoli and A.A. Balkema, Rotterdam, 6–9 October, 139–46.

Barazzetti, L., F. Banfi, R. Brumana, G. Gusmeroli, M. Previtali and G. Schiantarelli. 2015. Cloud-to-BIM-to-FEM: Structural simulation with accurate historic BIM from laser scans. *Simulation Modelling Practice and Theory* 57:71–87.

Cabaleiro, M., B. Riveiro, P. Arias, J.C. Caamaño and J.A. Vilán. 2014. Automatic 3D modelling of metal frame connections from LiDAR data for structural engineering purposes. *ISPRS Journal of Photogrammetry and Remote Sensing* 96:47–56.

Cereijo, J., S. Lagüela, D. Roca, J. Martínez and H. Lorenzo. 2014. Automatic detection of thermal bridges in thermographic building inspections. In *Proceedings of the 40th IAHS World Congress on Housing*, Funchal, 16–19 December.

Daniels, D.J. 2004. *Ground Penetrating Radar*. Institution of Electrical Engineering, London.

Hamrelius, T. 1992. Accurate temperature measurement in thermography. *QIRT* 27:39–45.

Laefer, D.F., L. Truong-Hong and M. Fitzgerald. 2011. Processing of terrestrial laser scanning point cloud data for computational modelling of building facades. *Recent Patents on Computer Science* 4:16–29.

Lagüela, S., J. Armesto, P. Arias and J. Herráez. 2012c. Automation of thermographic 3D modeling through image fusion and image matching techniques. *Automation in Construction* 27:24–31.

Lagüela, S., L. Díaz-Vilariño, J. Martínez and J. Armesto. 2013. Automatic thermographic and RGB texture of as-built BIM for energy rehabilitation purposes. *Automation in Construction* 31:230–240.

Lagüela, S., H. González-Jorge, J. Armesto and P. Arias. 2011. Calibration and verification of thermographic cameras for geometric measurements. *Infrared Physics & Technology* 54:92–99.

Lagüela, S., H. González-Jorge, J. Armesto and J. Herráez. 2012a. High performance grid for the metric calibration of thermographic cameras. *Measurement Science & Technology* 23:015402.

Lagüela, S., M. Solla, J. Armesto and H. González-Jorge. 2012b. Comparison of infrared thermography with ground-penetrating radar for the non-destructive evaluation of historic masonry bridges. In *Proceedings of the 11th International Conference on Quantitative InfraRed Thermography*, Naples, 11–14 June.

Lewis, D., H. Goller, and C. Teates. 1973. Apparent temperature degradation in thermograms of human anatomy viewed obliquely, *Radiology*, 106(1), 95–99.

Lubowiecka, I., P. Arias, B. Riveiro and M. Solla. 2011. Multidisciplinary approach to the assessment of historic structures based on the case of a masonry bridge in Galicia (Spain). *Computers and Structures* 89:1615–27.

Luo, W., J. Wu, J. Peng and B. Zhang. 2014. Influence of surface treatment of components on thermal radiation performance in infrared optical systems. In *Proceedings of the 7th International Symposium on Advance Optical Manufacturing and Testing Technologies: Large Mirrors and Telescopes*, ed. W. Jiang, M.K. Cho and F. Wu, Harbin, China, 26–29 April.

Martín, S., I. Cañas and I. González. 2004. Thermographic survey of two rural buildings in Spain. *Energy and Buildings* 34:515–23.

Meola, C. and G.M. Carlomagno. 2004. Recent advances in the use of infrared thermography. *Measurement Science and Technology* 15:R27–R58.

Olhoeft, G.R. 2000. Maximizing the information return from ground penetrating radar. *Journal of Applied Geophysics* 43:175–87.

Puente, I., M. Solla, H. González-Jorge and P. Arias. 2015. NDT documentation and evaluation of the Roman bridge of Lugo using GPR and mobile and static LiDAR. *Journal of Performance of Constructed Facilities* 29(1).

Riveiro, B., M.J. DeJong and B. Conde. 2016. Automated processing of large point clouds for structural health monitoring of masonry arch bridges. *Automation in Construction* 72:258–68.

Riveiro, B., P.B. Lourenço, D.V. Oliveira, H. González-Jorge and P. Arias. 2015. Automatic morphologic analysis of quasi-periodic masonry walls from LiDAR. *Computer-Aided Civil and Infrastructure Engineering* 31(4):305–19.

Riveiro, B., P. Morer, P. Arias and I. de Arteaga. 2011. Terrestrial laser scanning and limit analysis of masonry arch bridges. *Construction and Building Materials* 25(4):1726–35.

Riveiro, B., M. Solla, I. de Arteaga, P. Arias and P. Morer. 2013. A novel approach to evaluate masonry arch stability on the basis of limit analysis theory and non-destructive geometric characterization. *Automation in Construction* 31:140–48.

Rodríguez-Martín, M., S. Lagüela, D. González-Aguilera and P. Rodríguez-Gonzálvez. 2016. Applications of infrared thermography to the analysis of Steel welded junctions. *Advances in Materials Science Research* 26.

Rosina, E. and J. Spodek. 2003. Using infrared thermography to detect moisture in historic masonry: a case study in Indiana. *APT Bulletin*, 34(1): 11–16.

Sandmeier, K.J. 2016. ReflexW software. http://www.sandmeier-geo.de (accessed May 3, 2016).

Solla, M., R. Asorey-Cacheda, X. Núñez-Nieto and B. Conde-Carnero. 2016a. Evaluation of historical bridges through recreation of GPR models with the FDTD algorithm. *NDT&E International* 77:19–27.

Solla, M., J.C. Caamaño, B. Riveiro and P. Arias. 2012b. A novel methodology for the structural assessment of stone arches based on geometric data by the integration of photogrammetry and ground-penetration radar. *Engineering Structures* 35:296–306.

Solla, M., H. Lorenzo, F.I. Rial and A. Novo. 2010. GPR assessment of the medieval arch bridge of San Antón, Galicia, Spain. *Archaeological Prospection* 17:223–32.

Solla, M., H. Lorenzo, F.I. Rial and A. Novo. 2011a. GPR evaluation of the Roman masonry arch bridge of Lugo (Spain). *NDT&E International* 44:8–12.

Solla, M., H. Lorenzo, F.I. Rial and A. Novo. 2012a. Ground-penetrating radar for the structural evaluation of masonry bridges: Results and interpretational tools. *Construction and Building Materials* 29:458–65.

Solla, M., H. Lorenzo, B. Riveiro and F.I. Rial. 2011b. Non-destructive methodologies in the assessment of the masonry arch bridge of Traba, Spain. *Engineering Failure Analysis* 18(3):828–35.

Solla, M., B. Riveiro, B. Conde, P. Arias and H. Lorenzo. 2016b. The potential of the TLS and GPR methods for the SHM of masonry arch bridges. In *Proceedings of the 8th European Workshop on Structural Health Monitoring*, 5–8 July, Bilbao, Spain.

Solla, M., B. Riveiro, H. Lorenzo and J. Armesto. 2014. Ancient stone bridge surveying by ground-penetrating radar and numerical modeling methods. *Journal of Bridge Engineering* 19(1):110–19.

Stavroulaki, M.E., B. Riveiro, G.A. Drosopoulos, M. Solla, P. Koutsianitis and G.E. Stavroulakis. 2016. Modelling and strength evaluation of masonry bridges using terrestrial photogrammetry and finite elements. *Advances in Engineering Software* 101:136–48.

Tang, P., D. Huber, B. Akinci, R. Lipman, A. Lytle. 2010. Automatic reconstruction of as-built building information models from laser-scanned point clouds: A review of related techniques. *Automation in Construction* 19:829–43.

Truong-Hong, L., D.F. Laefer, T. Hinks and H. Carr. 2013. Combining an angle criterion with voxelization and the flying voxel method in reconstructing building models from LiDAR data. *Computer-Aided Civil and Infrastructure Engineering* 28(2):112–29.

Vosselman, G. and H.G. Maas. 2010. *Airborne and Terrestrial Laser Scanning*. Whittles Publishing, Dunbeath, UK.

Yang, L. and Y. Li. 2009. City ventilation of Hong Kong at no-wind conditions. *Atmospheric Environment* 43:3111–21.

13 Ground-Penetrating Radar for Archaeology and Cultural-Heritage Diagnostics

Activities Carried Out in COST Action TU1208

Lara Pajewski, Mercedes Solla and Melda Küçükdemirci

CONTENTS

13.1 INTRODUCTION

This chapter aims to offer a brief overview on the most interesting activities fulfilled by members of the COST (European COoperation in Science and Technology) Action TU1208: Civil Engineering Applications of Ground-Penetrating Radar, concerning the use of ground-penetrating radar (GPR) and complementary nondestructive testing (NDT) techniques in archaeology and cultural heritage diagnostics.

GPR is a safe, effective and nondestructive technique that uses electromagnetic waves to provide high-resolution images of the subsurface, or to assess the inner status of a structure (Benedetto and Pajewski, 2015). In archaeological studies, digitisation and the management of cultural heritage, GPR is especially useful. It can be successfully employed to discover and map buried archaeological artefacts (Santos-Assunçao et al., 2016); to inspect ancient buildings (Gonzalez-Drigo et al., 2015), bridges (Solla et al., 2014), columns (Pérez-Gracia et al., 2013) and statues

(Dimitriadis, 2015); to thoroughly check frescoes (Pieraccini et al., 2005), mosaics (Moropoulou et al., 2012) and decorations (Cupa et al., 2012); and to study the internal conditions of several other objects of historical value. The use of GPR is not limited to the investigation of man-made structures. Indeed, this technique can also be exploited for the inspection of natural structures of historical, geological, biological or landscape conservation value, such as the trunks and roots of veteran trees (Jezova et al., 2016), glaciers (Helfricht et al., 2014), craters and caverns (Gosar and Čeru, 2016), fossil beds (Lanzarone et al., 2016), sand dunes (Van Dam, 2012) and more (Vargemezis et al., 2014).

COST is the longest-running European framework supporting cooperation among scientists and researchers across Europe and beyond. Founded in 1971, it is currently integrated in the Horizon 2020 programme. It contributes to reducing the fragmentation in European research investment, building the European Research Area (ERA) and opening it to worldwide cooperation. It also aims at constituting a 'bridge' towards the scientific communities of emerging countries, increasing the mobility of researchers across Europe and fostering the establishment of excellence in key scientific domains. Gender balance, early-career investigators and inclusiveness are strategic priorities of the COST programme. COST does not fund research itself, but it provides support for activities carried out within Actions: these are bottom-up science and technology networks, centred around nationally funded research projects, with a four-year duration and a minimum participation of five countries. The Actions are active through a range of networking tools, such as meetings, workshops, conferences, training schools, short-term scientific missions and dissemination activities; they are open to researchers and experts from universities, public and private research institutions, nongovernmental organisations, industry and small and medium-sized enterprises.

The Action TU1208 started in April 2013 and ends in October 2017. It involves almost 150 partner institutes from 28 COST countries (Austria, Belgium, Croatia, Czech Republic, Denmark, Estonia, Finland, France, former Yugoslav Republic of Macedonia, Germany, Greece, Ireland, Italy, Latvia, Malta, the Netherlands, Norway, Poland, Portugal, Romania, Serbia, Slovakia, Slovenia, Spain, Sweden, Switzerland, Turkey and the United Kingdom), a COST cooperating state (Israel), 6 COST near neighbour countries (Albania, Armenia, Egypt, Jordan, Russia, Ukraine) and 6 COST international partner countries (Australia, Colombia, Hong Kong Special Administrative Region of the People's Republic of China, The Philippines, Rwanda, the United States). University researchers, software developers, civil and electronic engineers, archaeologists, geophysics experts, NDT equipment designers and producers, end users from private companies and stakeholders from public agencies are participating in the Action. The scientific structure of the Action includes four working groups (WGs). WG1 focuses on the design of novel GPR instrumentation. WG2 deals with the development of guidelines for the surveying of transport infrastructures and buildings, and for the sensing of underground utilities and voids. WG3 studies electromagnetic forward and inverse methods for the solution of near-field scattering problems by buried structures and data processing techniques. WG4 is concerned with applications of GPR outside of the civil engineering field and the integration of GPR with other technologies. The use of GPR for archaeological prospection and cultural heritage diagnostics is studied in WG4.

Section 13.2 focuses on the use of GPR for archaeological prospection. In this field, a wider application of GPR and complementary noninvasive investigation techniques needs to be fostered for the identification and visualisation of buried archaeological heritage. Archaeological excavation provides unparalleled information about the past, but it is a costly process and, by its very nature, is destructive to the site under study. By contrast, subsurface mapping methods can cover wide areas while leaving the site undisturbed. A couple of decades ago, these methods represented 'only' a means of finding sites; nowadays, they provide more information and offer a good overview of past human activity and buried structures of historical interest in a surveyed area. Considering the methodological and technological development that archaeological fieldwork has undergone during the past century, we can expect that future archaeologists will have at their disposal scientific possibilities that today are not even imaginable. Thus, in our opinion, important sites should be studied as much as possible by using noninvasive approaches and they should be kept intact for future generations; when digging is useful or necessary, subsurface mapping methods should be exploited to select the highest potential areas for targeted scientific excavations.

Still concerning archaeological prospection, it is worth stressing that, until very recently, subsurface mapping methods were suitable to be used 'only' for the investigation of individual archaeological sites or to explore limited areas with a known high potential of hosting structures of archaeological interest. In the last 10 years, thanks to technological advancements in GPR and other near-surface geophysical approaches, the sensitivity and resolution of these methods have increased and data can be acquired at a faster rate. Therefore, instead of mapping the remains of individual constructions, it is now possible to explore entire ancient towns and landscapes, covering several square kilometres rather than hectares (Neubauer et al., 2014; De Smedt and Trinks, 2014). This implies that a higher number of formerly unknown and otherwise invisible structures can be discovered; more complete information about a region of historical interest can be collected; it is possible to fill the gap between different buildings or ruins and study their relations; and the significance of a monument can be much better appreciated, because a deep understanding of the landscape around it can be achieved. The combined application of high-resolution prospection methods at the scale of landscapes, as well as their integrated interpretation, is still an uncommon approach and should be strongly encouraged because it is very promising and leads to a new way to carry out archaeological studies.

Section 13.3 deals with the use of GPR for the assessment of monuments, historical buildings, bridges and statues. Members of COST Action TU1208 have employed GPR, eventually combined with other nondestructive approaches, to inspect numerous structures of high historical value and of different time periods, ranging from the thirteenth century BC to the modernist period.

It is interesting to underline the huge importance, in these areas of application of the GPR technique, of modelling methods for the electromagnetic simulation of complex scenarios. Radargrams do not usually have a direct resemblance to the subsurface or to the structures over which GPR is used. Various factors, including the innate design of the survey equipment and the complexity of electromagnetic propagation in the scenario, can disguise complex structures recorded on GPR profiles.

Radar data need to be properly processed and interpreted to obtain geometrical and physical information about the surveyed structure or area. In that respect, the simulation of GPR scenarios is of primary importance in understanding how target structures get translated into radargrams. Electromagnetic models significantly aid in the interpretation of complex data sets, facilitate object localisation and shape the reconstruction and estimation of geophysical parameters; they help to find out unknown details about the investigated structure or subsurface. Additionally, an electromagnetic simulator can be employed before a survey, to support the choice of the most correct GPR equipment. TU1208 members have developed and released interesting open source or freeware tools for the electromagnetic modelling of GPR scenarios (Pirrone and Pajewski, 2015; Warren et al., 2016a, b), the most significant being a new version of the software gprMax, which implements the finite-difference time-domain (FDTD) technique. Finally, in Section 13.4 conclusions are drawn.

13.2 ARCHAEOLOGICAL PROSPECTION USING GROUND-PENETRATING RADAR

Archaeological features and materials are hidden beneath the surface as a consequence of natural effects and human activity over the years. In the past four decades, various geophysical methods have been used for archaeological prospection. GPR surveys started as small-scale approaches to detect buried features and identify their main geometric and physical properties. As mentioned in Section 13.1, nowadays a GPR investigation can easily cover wide areas and provide high-resolution information of entire landscapes, thus helping in understanding cultural and social relations and historical evolutions.

In this frame, two significant projects carried out by scientists participating in COST Action TU1208 need to be mentioned, namely the 'Carnuntum Project' (Neubauer et al., 2014; Valdelomar et al., 2015) and the 'Stonehenge Hidden Landscape Project' (Neubauer et al., 2014; De Smedt and Trinks, 2014).

The site of the Roman town of Carnuntum is located about 30 km south-east of Vienna, Austria, and is mostly covered with agriculturally used fields. It extends over an area of about 10 km^2. The whole area of Carnuntum, comprising a civil town, as well as a military camp and settlement, has been mapped in great detail. Magnetometer systems for large-scale prospection have been employed; earth-resistance measurements have been performed and remote sensing techniques have been used; the subsurface has been imaged with different single-channel and multichannel GPR systems. Extraordinary results have been obtained, including the unique discovery of the school of gladiators. Such a structure was first mapped with a single-channel 900 MHz GPR system, and subsequently investigated in great detail using a multichannel 400 MHz array with only 8 cm crossline spacing. More information is available at http://lbi-archpro.org/cs/carnuntum/.

As is well known, Stonehenge (UK) occupies one of the richest archaeological landscapes in the world, where archaeological and antiquarian research has been intensively carried out over centuries. The aim of the 'Stonehenge Hidden Landscape Project' was to address gaps in our knowledge and advance the understanding of the Stonehenge landscape by conducting a cutting-edge geophysical and remote sensing survey on an unprecedented scale. Hundreds of new features were discovered and a detailed archaeological digital map of the Stonehenge landscape was produced. More information about the project and the achieved results can be found at http://lbi-archpro.org/cs/stonehenge/.

In archaeology, researchers recommend applying at least two complementary non-destructive methods to obtain complete information about the investigated area. For many years, qualitative data integration techniques were quite favoured. Nowadays, quantitative data integration and fusion techniques are preferred and developed: this is a fascinating and promising field where more research is needed.

As for the use of GPR for the investigation of individual archaeological sites, among the most interesting studies carried out in the framework of TU1208 is the GPR evaluation of the Tholos Acharnon Tomb (Athens, Greece), an ancient Mycenaean monument dating back to the thirteenth century BC (Santos-Assunçao et al., 2016). This is a circular and buried structure, made with irregular stones. The curved rough surfaces of the tomb represented a challenge for data acquisition and interpretation. The main objectives of the survey were to identify hidden elements inside the walls, study the arrangement of the stones and detect damaged zones. During data acquisition, a laser scan was used to record the position and path of each radargram. The most interesting outcome of the investigation was the three-dimensional description of the internal structure of the tomb walls; furthermore, zones where the GPR signal was highly attenuated were identified, probably due to a high salt content.

Another interesting study was carried out by Action members from the Université catholique de Louvain (Belgium), in the Villa Gallo-Romaine de Mageroy, near Habay-La-Vieille (Belgium). The villa's origins date back to the first century AD and the site has been under investigation for more than 20 years by the Groupe d'Archéologie de Habay. By using GPR, new Mageroy secrets have been revealed: a new edifice to the villa and a drainage network have been unveiled. More information about the project can be found at http://sites.uclouvain.be/gprlouvain/mageroy.html.

The Carthusian monastery of Bourgfontaine, north-east of Paris (France), was established between 1323 and 1325 and hosted 24 brothers living in separate cells. Bourgfontaine survives today as a monumental ruin and a working farm. Here, Action members carried out a GPR survey over a rectangular area of 30 m × 50 m (Saintenoy et al., 2014). Time slices at different depths provided clear evidence of buried architectural structures. The performed superposition of the surveyed zone with the extrapolated plan was particularly remarkable, allowing for a clear identification of three monastic cells as well as the foundations of the great cloister alley. Based on the GPR results, targeted excavations were planned and carried out.

13.3 GROUND-PENETRATING RADAR ASSESSMENT OF MONUMENTS, HISTORICAL BUILDINGS, BRIDGES AND STATUES

Members of COST Action TU1208 performed a number of cutting-edge GPR investigations on monuments, historical buildings and bridges. Moreover, some studies were concerned with the GPR assessment of statues.

Significant results were obtained in St. Mary's Cathedral of Mallorca (Spain), a Catalan Gothic-style building dating back to the fourteenth century (Pérez-Gracia et al., 2013; Santos-Assuncao et al., 2014). Using GPR combined with seismic tomography, the main activities were aimed at assessing the internal conditions of a series of different columns. Images of the internal structure of the columns were obtained and cracks not visible from the outside were identified. More columns were assessed in the twentieth-century Hospital de la Santa Creu i Sant Pau, built in the modernist style in Barcelona, Spain (Gonzalez-Drigo et al., 2015; Santos-Assunçao et al., 2015). These results are of particular interest because the application of GPR to vertical structural elements is still rather uncommon. Electromagnetic models of columns were also implemented using the gprMax simulator, and very good agreement between experimental and synthetic radargrams was obtained (Pérez-Gracia and Pajewski, 2017). It is worth mentioning that GPR technology and the methodology developed for the assessment of columns also has application in the field of tree trunk investigation (Jezova et al., 2016).

In 2015, the Action organised a training school in Cracow (Poland) on the 'Applications of Ground-Penetrating Radar in Urban Areas: The Sensitive Case of Historical Cities'. An insight into the challenges, advantages and potential of GPR prospection in historical cities was presented. Experimental sessions were held in the famous St. Leonard's Crypt, located under the Wawel Cathedral of St. Stanislaus BM and St. Wenceslaus M and built in the years 1090–1117. The floor of the crypt was surveyed to obtain an image of the tomb of Bishop Maurus, verify whether further cavities were present and collect information about the subsurface of the crypt. GPR scans were taken on a 20 × 20 cm grid. Subsequently, a smaller area of interest was chosen, where further data were collected on a 10 × 10 cm grid. The main finding was that the tomb of Bishop Maurus had been moved, with respect to the inscription placed in the middle of the crypt which is supposed to indicate its position. Moreover, it was possible to detect the presence of another large cavity and estimate its size (Benedetto et al., 2016).

The monuments and historical buildings assessed using GPR in the framework of TU1208 also include: the Midas monument in the Eskişehir province of Turkey, dating back to the seventh or sixth century BC (Kadioglu et al., 2015); the Sultan Alp Arslan Tomb of Merv in Turkestan, built in the eleventh century (Kadioglu and Kadioglu, 2016); the sixteenth-century St. John's Co-Cathedral in Valletta (Malta) (Persico et al., 2016a); the Jesuits' church and botanical gardens in Valletta (Malta) (Persico et al., 2016b); and the seventeenth-century Keciova Mosque of Casbah-Algiers in Algeria (Kadioglu et al., 2014). For brevity, details on those studies cannot be provided here, but all the information can be found in the cited references.

Historical bridges represent an important value from an architectural, historical, economic, symbolic and aesthetic point of view. They often characterise the landscape of a region and represent an integral part of its traditional architectonic heritage. Most bridges that remain from Roman and medieval times are stone masonry arch bridges that have lost their original utility and now have different functionalities from those they were originally designated. Some of them are still part of the road infrastructure and their use causes deterioration that demands constant evaluation and conservation activities, due to the strong loads and intense vibrations they have to support. When conservation stops, ruin and destruction soon set in. This has posed an important task for engineers and scientists, in terms of determining the state of conservation of historical bridges and providing information about their preservation and restoration using nondestructive methods that do not change the historical character of the structure.

GPR is one of the most recommended methods in bridge inspection. Valuable previously unknown information has been revealed about important historical bridges in Spain that were studied by Action members, such as the presence of hidden arches (Solla et al., 2014), the geometry of ancient profiles (Puente et al., 2015) and evidence of restoration and reconstruction tasks (Solla et al., 2016). Notable information was also revealed from a civil structural point of view. Thus, it was possible to detect and map different building materials or inhomogeneity in the stonework that affects the structural stability (Puente et al., 2015), ring-stone thickness and moist areas (Solla et al., 2013), the existence of a cavity in a pier and the nature of bridge foundations.

In complex and heterogeneous structures, such as masonry arch bridges, the interpretation and analysis of GPR data are often complicated. Numerical modelling has represented an affective interpretational tool when used in conjunction with experimental GPR campaigns. In Solla et al. (2016), the FDTD algorithm has been used for interpreting the complex pattern of reflections obtained in the field data due to the heterogeneity in the filling.

The combination of GPR with complementary noninvasive techniques turned out to be very useful in the assessment of historical masonry bridges. Different techniques can be employed in the evaluation and testing of masonry arch bridges, such as sonic/ultrasonic, electromagnetic and electrical methods as well as infrared thermography. Fauchard et al. (2013) obtained interesting information such as voids, moisture content and the heterogeneity in the filling. They employed two complementary geophysical methods: electrical resistivity tomography and GPR; moreover, photogrammetry was used to construct a three-dimensional model of the bridge which improved the interpretation of the field data. Solla et al. (2016) have shown the combined use of GPR, thermography and photogrammetry to evaluate the internal structure of historic bridges. The metric information obtained by the photogrammetric survey allowed the calculation of the average velocity of the radar wave for the filling and the masonry within the bridge, providing a better GPR data interpretation. The thermographic data produced allowed the characterisation of the inner materials while GPR data revealed the distribution of the layering and filling as well as identifying previously unknown construction details.

Finally, other authors have presented a multidisciplinary approach by means of laser scanning or photogrammetric methods and GPR as well as finite element analysis for the structural assessment of historical bridges (Stavroulaki et al., 2016). A complex external geometry of the structure is provided using the data collected by a laser scanner. The accuracy of the geometry obtained can be useful in performing an exhaustive interpretation of the GPR data, obtaining information about the composition of the structure. The resulting information is used to define a hypothesis for the numerical model and the finite element structural analysis, which provides the structural behaviour of the bridge. Other works have combined different geophysical methods with finite element method analysis to obtain complementary information on the internal state of the bridge. Bergamo et al. (2015) combined the use of thermography, seismic and GPR methods for an integral finite element analysis of historical masonry arch bridges. Russo (2016) employed sonic and GPR methods to analyse the structural behaviour of the Rialto Bridge in Venice (Italy).

With the progress in technology, lightweight GPR systems working above 2 GHz have been used, providing a resolution of a few tenths of a millimetre by means of specific processing. Consequently, new applications have become possible, such as the accurate assessment of the internal conditions of statues and architectural decorations. An interesting study carried out by TU1208 members was concerned with testing five marble statues from the 'Antikythera Shipwreck' collection, conserved in the National Archaeological Museum of Athens (Greece): 'Wrestler,' 'Hermes,' 'Helmeter,' 'Achilles' and 'Ulisse' (Dimitriadis, 2015). The Antikythera Shipwreck collection consists of underwater findings from the bottom of the sea near Antikythera Island. All statues suffered substantial erosion because of their exposure to seawater for more than 2000 years. Most of them were found in parts and were reassembled by using metal joints, but conservation records were not kept. The dual purpose of the project was to detect, using GPR, the presence of metallic joints in the statues as well as to identify hidden structural problems. The main challenge of the investigation was represented by the need to scan objects with undulated and rough surfaces. A GPR system with a 2.4 GHz antenna was used. In most cases, it was possible to successfully identify and locate the metal targets.

13.4 CONCLUSIONS

This chapter provided a brief overview on the activities fulfilled by members of the COST Action TU1208: Civil Engineering Applications of Ground-Penetrating Radar, concerning the use of GPR in archaeology and cultural heritage diagnostics. A rapid review of large-scale inspections performed in Carnuntum and Stonehenge was provided, as well as a review of surveys carried out at further archaeological sites of more limited extent. Case studies were mentioned, where structures of great historical value and from different time periods were prospected, ranging from the thirteenth century BC to the modernist period and including tombs, monuments, cathedrals, buildings and bridges. Information about GPR assessment of statues was also provided.

ACKNOWLEDGEMENT

This chapter is a contribution to COST (European COoperation in Science and Technology) Action TU1208: Civil Engineering Applications of Ground-Penetrating Radar (www.cost.eu, www.GPRadar.eu). The authors thank COST for funding and supporting Action TU1208.

REFERENCES

Benedetto, A. and Pajewski, L. 2015. *Civil Engineering Applications of Ground Penetrating Radar*. Springer, Book Series, Springer Transactions in Civil and Environmental Engineering, 385 pp., doi:10.1007/978-3-319-04813-0.

Benedetto, A., Pajewski, L., Dimitriadis, K., Avlonitou, P., Konstantakis, Y., Musiela, M., Mitka, B., Lambot, S. and Zakowska, L. 2016. Ground-penetrating radar investigation of St. Leonard's crypt under the Wawel Cathedral (Cracow, Poland): COST Action TU1208. *Geophysical Research Abstracts* 18: EGU2016-17965-1.

Bergamo, O., Campione, G., Donadello, S. and Russo, G. 2015. In-situ NDT testing procedure as an integral part of failure analysis of historical masonry arch bridges. *Engineering Failure Analysis* 57: 31–55.

Cupa, A., Felczak, M., Kurkiewicz, T., Lament, A., Rogal, R., Rogós, J., Szroeder, P. and Więcek, B. 2012. Thermovision and ground penetrating radar for investigation of architectural decorations: Potential and limitations. *Proceedings of the 11th International Conference on Quantitative InfraRed Thermography*, 11–14 June, Naples, Italy: 1–6.

De Smedt, P. and Trinks, I. 2014. STSM 4: Reconstructing prehistoric environments at Stonehenge with multiple electromagnetic survey methods. *Short Term Scientific Missions and Training Schools: Year 1*, Aracne, Ariccia, ISBN 978-88-548-7225-7: 64–76.

Dimitriadis, K. 2015. GPR in the preservation of cultural heritage. *Proceedings of Third General Meeting, COST Action TU1208*, London, UK, March. Aracne, Ariccia. Book series: COST Action TU1208, Rome, Italy. ISBN 978-88-548-8486-1.

Fauchard, C., Antoine, R., Bretar, F., Lacogne, J., Fargier, Y., Maisonnave, C., Guilbert, V. et al. 2013. Assessment of an ancient bridge combining geophysical and advanced photogrammetric methods: Application to the Pont De Coq, France. *Journal of Applied Geophysics* 98:100–112.

Gonzalez-Drigo, R., Perez-Gracia, V. and Santos-Assuncao, S. 2015. GPR assessment of the modernist Santa Creu i Sant Pau Hospital in Barcelona: GPR application in cultural heritage. *Proceedings of the International Workshop on Advanced Ground Penetrating Radar (IWAGPR)*, 7–10 July, Florence, Italy: 1–4, doi: 10.1109/IWAGPR.2015.7292658.

Gosar, A. and Čeru, T. 2016. Search for an artificially buried karst cave entrance using ground penetrating radar: A successful case of locating the S-19 cave in the Mt. Kanin massif (NW Slovenia). *International Journal of Speleology* 45(2): 135–147.

Helfricht, K., Kuhn, M., Keuschnig, M. and Heilig, A. 2014. Lidar snow cover studies on glaciers in the Ötztal Alps (Austria): Comparison with snow depths calculated from GPR measurements. *Cryosphere* 8(1): 41–57.

Jezova, J., Mertens, L. and Lambot, S. 2016. Ground-penetrating radar for observing tree trunks and other cylindrical objects. *Construction and Building Materials* 123: 214–225, doi:10.1016/j.conbuildmat.2016.07.005.

Kadioglu, S. and Kadioglu, Y. K. 2016. Investigation of the Sultan Alp Arslan tomb with geophysical methods, in the historical Merv city (Turkestan). *Geophysical Research Abstracts, European Geosciences Union (EGU) General Assembly 2016*, 17–22 April, Vienna, Austria: article ID EGU2016-18494.

Kadioglu, S., Kadioglu, K. Y. and Akyol, A. A. 2015. Picturing monuments and cultural heritages with ground penetrating radar method including its half bird's eye view visualization. *Proceedings of the 8th International Workshop on Advanced Ground Penetrating Radar (IWAGPR)*, 7–10 July, Florence, Italy: 1–4, doi: 10.1109/IWAGPR.2015.7292684.

Kadioglu, S., Kadioglu, Y. K. and Deniz, K. 2014. Ground penetrating radar and micro Raman spectroscopy in Keciova Mosque Casbah-Algiers (Algeria). *Proceedings of the 15th International Conference on Ground Penetrating Radar (GPR 2014)*, 30 June–4 July, Brussels, Belgium: 50–53, doi: 10.1109/ICGPR.2014.6970383.

Lanzarone, P., Garrison, E., Bobe, R. and Getahun, A. 2016. Examining fluvial stratigraphic architecture using ground-penetrating radar at the Fanta Stream fossil and archaeological site, central Ethiopia. *Geoarchaeology* 31(6): 577–591.

Moropoulou, A., Labropoulos, K. C. and Katsiotis, N. S. 2012. Assessing the preservation state and revealing plastered mosaics in Hagia Sophia using ground penetrating radar. *Journal of Materials Science & Engineering A* 2(2): 183–189.

Neubauer, W., Gug, C., Scholz, M., Verhoeven, G., Trinks, I., Locker, K., Doneus, M., Saey, T. O. and Van Meirvenne, M. 2014. The discovery of the school of gladiators at Carnuntum, Austria. *Antiquity* 88: 173–190.

Pérez-Gracia, V., Caselles, J. O., Clapés, J., Martinez, G. and Osorio, R. 2013. Non-destructive analysis in cultural heritage buildings: Evaluating the Mallorca cathedral supporting structures. *Non Destructive Testing and Evaluation International* 59: 40–47.

Pérez-Gracia, V. and Pajewski, L. 2017. STSM 8: An educational package to teach GPR in the university. *Short Term Scientific Missions: Year* 3: 199–209.

Persico, R., Matera, L., D'Amico, S., Borg, R. P. and Galea, P. 2016a. Integrated GPR and passive seismic investigations at cultural heritage sites: Case studies in Malta. *Proceedings of the 16th International Conference on Ground Penetrating Radar (GPR 2016)*, Hong Kong, 13–16 June: 1–5, doi: 10.1109/ICGPR.2016.7572608.

Persico, R., Pajewski, L., D'Amico, S. and Pérez-Gracia, V. 2016b. Ground-penetrating radar prospection at the Jesuits' church in Valletta, Malta. *Proceedings of the 22nd European Meeting of Environmental and Engineering Geophysics: Near Surface Geoscience 2016*, 4–8 September, Barcelona, Spain: 1–4, doi: 10.3997/2214-4609.201602031.

Pieraccini, M., Mecatti, D., Luzi, G., Seracini, M., Pinelli, G. and Atzeni, C. 2005. Non-contact intrawall penetrating radar for heritage survey: The search of the 'Battle of Anghiari' by Leonardo da Vinci. *Non Destructive Testing and Evaluation International* 38: 151–157.

Pirrone, D. and Pajewski, L. 2015. E2GPR; Edit your geometry, execute GprMax2D and plot the results!. *Proceedings of the 2015 IEEE 15th Mediterranean Microwave Symposium*, 30 November–2 December, Lecce, Italy: 1–4, doi: 10.1109/MMS.2015.7375414.

Puente, I., Solla, M., González-Jorge, H. and Arias, P. 2015. NDT documentation and evaluation of the Roman bridge of Lugo using GPR and mobile and static LiDAR. *Journal of Performance of Constructed Facilities* 29(1).

Russo, S. 2016. Integrated assessment of monumental structures through ambient vibrations and ND tests: The case of Rialto Bridge. *Journal of Cultural Heritage* 19: 402–414.

Saintenoy, A., Rejiba, F., Leger, E., Bonde, S. and Maines, C. 2014. Ground penetrating radar prospection at the charterhouse of Bourgfontaine. *Proceedings of the 15th International Conference on Ground Penetrating Radar (GPR 2014)*, Brussels, Belgium, 30 June–4 July: 83–86, doi: 10.1109/ICGPR.2014.6970390.

Santos-Assunçao, S., Dimitriadis, K., Konstantakis, Y., Perez-Gracia, V., Anagnostopoulou, E. and Gonzalez-Drigo, R. 2016. Ground-penetrating radar evaluation of the ancient Mycenaean monument Tholos Acharnon tomb. *Near Surface Geophysics* 14(2): 197–205, doi:10.3997/1873-0604.2015030.

Santos-Assunçao, S., Gonzalez-Drigo, R. and Perez-Gracia, V. 2015. GPR assessment of columns in the pavilions and the church of the modernista building Sant Pau Hospital, in Barcelona. *Proceedings of Near Surface Geoscience 2015: 21st European Meeting of Environmental and Engineering Geophysics*, 6–10 September, Turin, Italy: 1–5, doi: 10.3997/2214-4609.201413674.

Santos-Asssunçao, S., Pérez-Gracia, V., González, R., Caselles, O., Clapés, J. and Salinas, V. 2014. Geophysical exploration of columns in historical heritage buildings. *Proceedings of the 15th International Conference on Ground Penetrating Radar (GPR 2014)*, Brussels, Belgium, 30 June–4 July: 97–102, doi: 10.1109/ICGPR.2014.6970393.

Solla, M., Asorey-Cacheda, R., Núñez-Nieto, X. and Conde-Carnero, B. 2016. Evaluation of historical bridges through recreation of GPR models with the FDTD algorithm. *NDT&E International* 77: 19–27.

Solla, M., Lagüela, S., Riveiro, B. and Lorenzo, H. 2013. Non-destructive testing for the analysis of moisture in the masonry arch bridge of Lubians (Spain). *Structural Control and Health Monitoring* 20: 1366–1376.

Solla, M., Riveiro, B., Lorenzo, H. and Armesto, J. 2014. Ancient stone bridge surveying by ground-penetrating radar and numerical modeling methods. *Journal of Bridge Engineering* 19(1): 110–119.

Stavroulaki, M. E., Riveiro, B., Drosopoulos, G. A., Solla, M., Koutsianitis, P. and Stavroulakis, G. E. 2016. Modelling and strength evaluation of masonry bridges using terrestrial photogrammetry and finite elements. *Advances in Engineering Software* 101: 136–148.

Valdelomar, J. T., Brandtner, J., Kucera, M., Wallner, M., Sandici, V. and Neubauer, W. 2015. 4D investigation of digital heritage: An interactive application for the auxiliary fortress of Carnuntum. *Proceedings of the 2015 Digital Heritage International Congress*, Digital Heritage 2015: Art. no. 7419457: 81–84.

Van Dam, R. L. 2012. Landform characterization using geophysics: Recent advances, applications, and emerging tools. *Geomorphology* 137(1): 57–73.

Vargemezis, G., Diamanti, N., Tsourlos, P. and Fikos, I. 2014. Electrical resistivity tomography and ground penetrating radar for locating buried petrified wood sites: A case study in the natural monument of the petrified forest of Evros, Greece. *Geophysical Research Abstracts|European Geosciences Union (EGU) General Assembly 2014*, 27 April–2 May, Vienna, Austria: article ID EGU2014-14298.

Warren, C., Giannopoulos, A., and Giannakis, I. 2016b. gprMax: Open source software to simulate electromagnetic wave propagation for ground penetrating radar. *Computer Physics Communications* 209: 163–170

Warren, C., Pajewski, L., Poljak, D., Ventura, A., Giannopoulos, A., and Sesnic, S. 2016a. A comparison of finite-difference, finite-integration and integral-equation methods in the time domain for modelling ground penetrating radar antennas. *Proceedings of the 16th International Conference on Ground Penetrating Radar (GPR 2016)*, Hong Kong, 13–16 June: 1–6, doi: 10.1109/ ICGPR.2016.7572676.

14 3D Reconstruction of the Roman Site *Aquis Querquennis* by Means of TLS and GPR Methods

Iván Puente, Javier Sanjurjo-Pinto, Mercedes Solla and Rafael Asorey-Cacheda

CONTENTS

14.1 INTRODUCTION

Cultural heritage is a fragile, nonrenewable resource that constitutes the basic record of past human activities, and as such, it should be protected (Vecco 2010). The protection and management of cultural heritage include materials characterisation, archaeological prospecting and structural stability analysis. Regarding archaeological surveys, the full characterisation of archaeological sites using conventional techniques takes years, while drilling and excavations, which are most frequently applied in archaeological assessments, are ground disturbing. Hence, in most cases, surveying is carried out using nondestructive testing (NDT) technologies, reflecting a societal concern for the maintenance and conservation of these sites (Lin et al. 2011).

From a point of view based on the three-dimensional (3D) representation of the complete as-built reality, the most commonly used technology is terrestrial laser scanning (TLS) (Yastikli 2007). However, buried structures cannot be traditionally assessed by visual and optical inspection. Alternative methods are therefore used for subsurface inspection, including geophysics such as magnetic, electrical and electromagnetic methods (Corsi et al. 2013).

In recent years, the use of TLS has become increasingly popular not only for the digital recording, management and documentation of cultural heritage items (Lerma et al. 2010, Puente et al. 2015) but also for the monitoring (González-Jorge et al. 2012, Puente et al. 2014a,b) and structural analysis of civil infrastructure (Riveiro et al. 2011), as this technology automatically provides very dense 3D points of an object's surface with high accuracy. Furthermore, fully integrated digital cameras in current TLS systems supply colour information to assist the process of documentation, together with photorealistic 3D models and digital orthoimages that can be easily generated.

In the matter of archaeogeophysics, one of the most noteworthy methods has been the ground-penetrating radar (GPR) method. It is a fast data acquisition technique that is commonly applied for high-resolution imaging in many archaeological applications (Goodman and Piro 2013). However, understanding the GPR data has been a long-term challenge among the nongeophysical community. The analysis of the radar signal and the interpretation of the measured data (2D GPR images) are not straightforward. The use of 3D GPR data processing and visualisation has aided in the acceptance of the use of GPR for archaeological prospecting (Grasmueck 2005). The use of 3D imaging techniques and processing software produces more realistic images of buried archaeological remains (Leckebusch 2003, Novo et al. 2014), which permits not only finding but also obtaining 3D reconstructions of buried structures (Núñez-Nieto et al. 2014).

This work presents the 3D reconstruction of the Roman site *Aquis Querquennis* (Bande, Spain). Both TLS and 3D GPR data acquisitions were carried out to provide high resolution images and 3D visualisation. The resulting 3D images are also georeferenced, which can be integrated into a geographic information system (GIS) environment. The hybrid outcomes derived from both methods can thus be merged into a single image, allowing for the reconstruction of the visible reality of the archaeological site while integrating the unexcavated structures detected in the subsurface.

14.2 ROMAN SITE *AQUIS QUERQUENNIS*

Aquis Querquennis was a Roman encampment located in Bande (Galicia, Spain), on the banks of the Limia River, occupied between the last quarter of the first century and the middle of the second century. It has been demonstrated that the site was conquered by a legion *septima Gemina* detachment, a mixed unit of cavalry and infantry. It was most probably established during the reign of Vespasian (69–79 AD). The encampment was probably built to monitor the building of the roads that connected *Bracara Augusta* (Braga, Portugal) and *Asturica Augusta* (Astorga, Spain). Some decades later, the encampment disappeared as such and at the same time, its structures were used as a quarry. The site grew until it became one of the most famous urban nucleus in interior *Gallaecia* (Galicia) in the middle of the second century. The first archaeological excavations were made in the 1920s and new studies were authorised in 1975, centred on the north-western area (Rodríguez 2016).

The settlement occupies 3 ha and its layout is classic: rectangular with two main orthogonal paths (Figure 14.1). It has a wall of softened corners, with a height of 3.20 m, and semi-cylindrical battlements separated 11 m from the interior constructions. There is a V-shaped moat outside the wall, 4 m deep and 3 m wide.

The encampment has four main doors, corresponding to the ends of the two aforementioned paths but only two have been excavated. Three barracks (*strigia*) have been identified, each one housing two *centuria* with their respective commands. They are composed of two rectangular wings faced around a courtyard (*compluvium*) with a cistern to collect rainwater. Every eight soldiers formed a *contubernium*, which was divided into a small front room (*arma*) where the soldiers kept their equipment and a rear room (*papilio*) for sleeping. In total, each *centuria* has 10 units of these characteristics, 6 in 1 wing, plus the guard quarter, and 4 in the opposite wing with a centurion's suite of rooms at the entrance. Gates and hollows are oriented to the south.

Moreover, there are two rectangular granaries raised on stone pillars and delimited by thick walls with buttresses, so the roofs are believed to be vaulted. A building has been found of almost square plant and rooms arranged around an *impluvium* (a sunken area designed to carry away rainwater). It is believed that it might have been a hospital (*valetudinarium*).

The central building was probably the headquarters. It has a dimension of 34.8 × 32.1 m, a vestibule flanked by ambulatories, a large central courtyard, a basilica accessed by three doors and a sacral-administrative area or official temple.

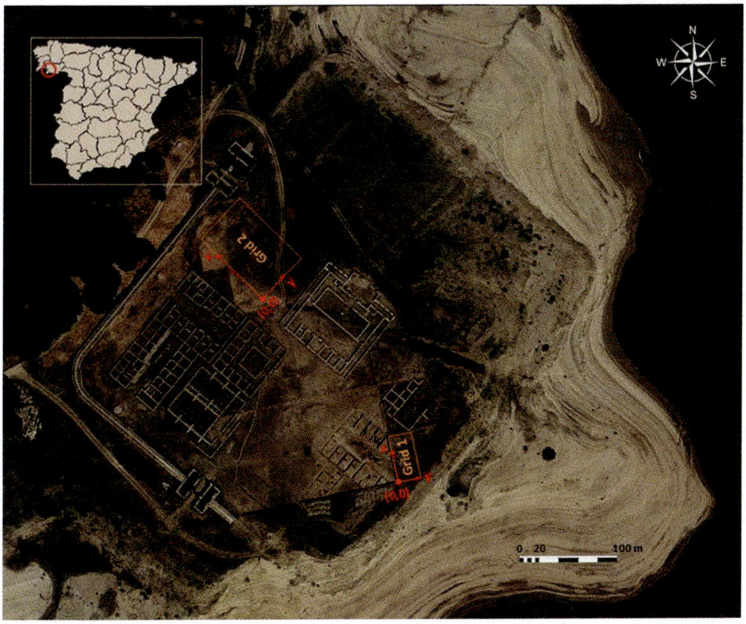

FIGURE 14.1 Situation of the Roman site *Aquis Querquennis* in Galicia (north-west of Spain) and grids surveyed with the GPR technique.

14.3 METHODS: DATA ACQUISITION AND PROCESSING

14.3.1 TLS Survey

A TLS measures the distance to an object that reflects the laser beam, with a field of view of 360º in the horizontal direction and up to 320° in the vertical direction. As a result, a very dense point cloud is obtained, containing the Cartesian coordinates of each point. Additionally, the intensity of the reflection is recorded and for those scanners with a red-green-blue (RGB) camera attached, the RGB value of the reflection is also captured.

The TLS survey was performed in the encampment using a Faro Focus^{3D} X 330 phase-shift scanner, with an integrated 8Mpix high dynamic range (HDR) camera to provide additional RGB information (Faro 2016) and a topographic tripod for its placement. Before the acquisition of the TLS data, a scan planning was developed to solve the location and the number of scans, the resolution, occlusions and the reference system. Because of the complexity and extension of the settlement, a total of 48 scan positions were performed with a base distance between 10 and 15 m covering all the encampment buildings and obtaining their corresponding point clouds with a spatial resolution of 6–7 mm/10 m distance. However, with three barracks divided into multiple rooms, some features of the terrain are not accessible in the TLS data due to occlusion, though the overall documentation process was not affected.

All the scans were then transformed into one common and known coordinate system (e.g. WGS-84), called the *registration process*. This point cloud alignment was done by a solid rigid transformation using common surfaces and supported by the iterative closest point (ICP) algorithm (Besl and McKay 1992), which focuses on reducing the distance between corresponding points in the point clouds. Since the registration had a large influence on the final quality of the resulting point cloud, accuracy of the order of a few millimetres was required. Finally, the raw laser point clouds were filtered and segmented to remove those points that were not part of the Roman site *Aquis Querquennis*.

14.3.2 GPR Survey

The GPR is a geophysical method based on the emission of electromagnetic pulses into the ground by a transmitting antenna. This signal is partly reflected when it encounters media with different dielectric properties and the produced reflections are recorded from a receiving antenna. The strength or intensity of the reflection is provided in terms of amplitude value. Amplitude is higher when the dielectric contrast between two different media is greater. By moving the antenna over the ground, a 2D image (radargram) is obtained, which are *XZ* graphic representations of the detected reflections. The *X* axis represents the antenna displacement along the survey line, and the *Z* axis represents the two-way travel time of the pulse emitted (in terms of nanoseconds). If the time required for the electromagnetic pulse to go from the transmitting antenna to the reflector into the ground and return to the receiving antenna is measured and the velocity of this pulse in the subsurface medium is known, then the position of the reflector can be determined.

The GPR survey was conducted using a RAMAC system from MALÅ Geosciences. A 500 MHz antenna was used, because this frequency provides proper vertical resolution, as well as sufficient penetration. The data acquisition was carried out using the common offset mode with the antenna polarisation perpendicular to the data collection direction; the acquisition parameters were 2 cm trace-distance interval and a total time window of 75 ns (512 samples). For trace-distance interval calculation while measuring the profile lengths, the GPR antennas were mounted on a survey cart with an encoder (odometer wheel).

3D GPR methodologies were performed in this work. Two grids of approximately 156 and 787 m^2 were designed to prospect different unexcavated areas of the Roman site (see Figure 14.1). To cover the entire grids, parallel 2D profiles were registered at regular intervals 20 cm apart resulting in a total of 41 and 123 profiles, respectively.

The data processing was carried out with a developed tool 'toGPRi'. This tool was developed in GNU Octave programming language and implements signal amplification and noise removal filters. The 2D measured data was processed by applying the following filtering sequence: time-zero correction, subtract mean trace, gain function with exponential and linear factors, manual gain, vertical and horizontal smoothing and background removal. Topographic corrections were also applied to compensate for variations in elevation along the profile line. Regarding the 3D data processing and visualisation, this tool allowed for the creation of a 3D model and subsequent *XY* images (raster layers) in depth. It also allowed the creation of an overlaid image to connect reflections at different levels and reconstruct the buried structures. All these 3D images are also georeferenced to the WGS-84 system that enables these results to be integrated into a GIS environment. Thus, the 3D GPR imaging and the orthoimages provided by TLS can be overlaid for an integrated interpretation of both visible and underground space.

14.4 RESULTS

14.4.1 Results of TLS Survey

As soon as this Roman encampment has been reconstructed from laser scanning data and the global point cloud has been obtained, many products and information applied to cultural heritage can be extracted. In this regard, the final product is not only a 3D surface model (see Figure 14.2c) for simple archaeological and architectural documentation and qualitative interpretation, but also the quantitative and accurate identification of geometric anomalies of structural elements.

With this approach, 2D orthoimages are highly attractive products as they combine accuracy and radiometric information, useful in providing both quantitative and qualitative results. Orthoimages were generated by applying an affine transformation to each of the planes of the triangular mesh obtained from the TLS point cloud data. The same mathematical model directly relates the pixel coordinates of the registered RGB values from the 8Mpix HDR camera to those of the orthoimages. Using these orthoimages, it is possible to carry out a dimensional analysis using computer-aided design (CAD) software, and for example, to accurately define the height of the structure, fronts, plans and customisable sets of horizontal and transversal sections (see Figure 14.2a and b). They can also be taken into account in pathology detection,

FIGURE 14.2 Results of the TLS survey: (a) orthoimage of the barrack at the western area of the Roman site, (b) orthoimage of the main door located in the south-western area and (c) 3D coloured point cloud of the same barrack as shown in (a).

for assessing the total surface affected by rock pathologies or measuring the size and orientation of rock fissures.

Lastly, light detection and ranging (LiDAR)-derived products can serve as valuable diagnostic tools for structural analysis, when their extracted parameters are the starting point for advanced studies on the statics and the dynamics of a structure.

The previous applications are undoubtedly of great help to managers, archaeologists and conservators in designing and planning future interventions on the site, as well as providing an accurate record of this rapidly deteriorating Roman site.

14.4.2 Results of GPR Prospection

Figure 14.3a presents the 3D overlaid image obtained in Grid 1 (see Figure 14.1), showing parts of some walls of a rectangular building composed of square rooms. The overlaid image was built considering the time slices (*XY* images) from 105 to

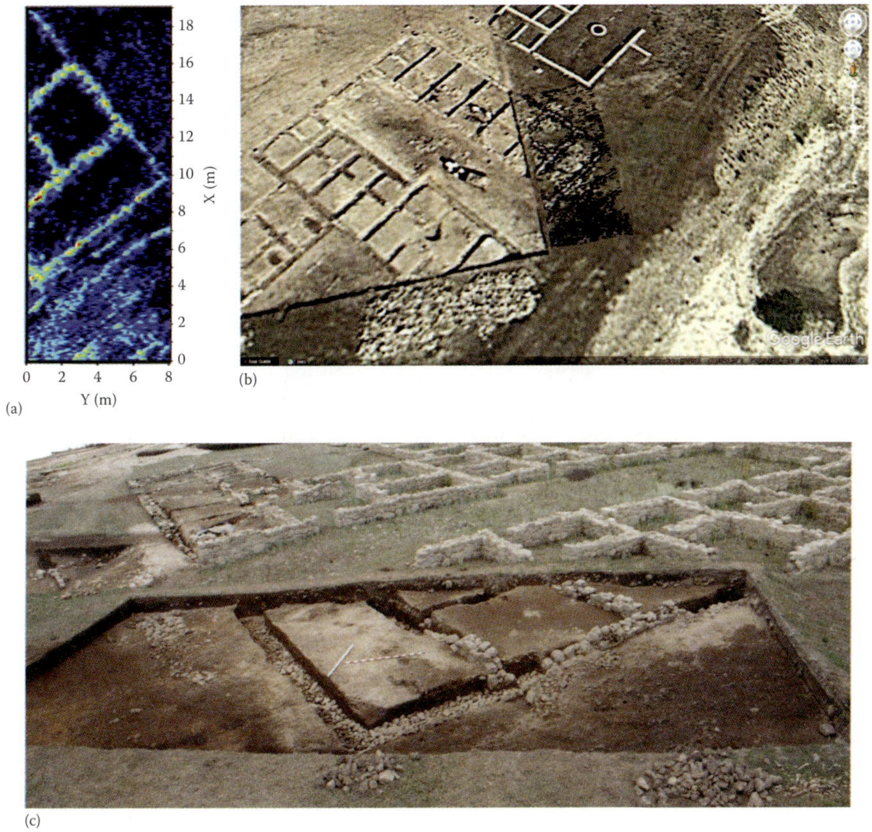

FIGURE 14.3 Results of the Grid 1 at the southern sector of the camp: (a) 3D overlaid image, (b) superposition of the 3D image with an aerial photograph of the site in Google Earth and (c) recent excavation.

140 cm in depth (with a precalibrated velocity of propagation of 11.6 cm/ns). Grid 1 has completed the barrack located at the southern sector of the encampment in Figure 14.1, and more specifically, it continues the unexcavated line of *contubernia* at the right wing of the barrack. The interpretation of the structures was performed within Google Earth by uploading the overlaid image and satellite images (Figure 14.3b). For these purposes, the developed tool 'toGPRi' allows exporting the raster images produced into a compressed KML file. Given these images, it was possible to achieve an integrated interpretation and to outline that this right wing of the barrack is composed of six *contubernia*, rectangles of 6 × 3 m, divided into two zones (*papilio* and *arma*) and presenting a guard quarter (8 × 3 m size) at the entrance of the barrack. It reveals a similar structure for the barrack at the northwestern sector, the one closest to the granaries. Finally, Figure 14.3c shows the discovered walls in Grid 1 after a recent excavation conducted in the 2016 campaign confirming such interpretation.

Figure 14.4 presents the 3D overlaid image obtained in Grid 2 (see Figure 14.1) from 70 to 150 cm in depth (with a velocity of propagation of 11.6 cm/ns). The results produced revealed the partial structure of a barrack, with two rectangular wings faced around a courtyard. The upper wing presents five *contubernia* of 6 × 3 m divided into two separate zones, *papilio* and *arma* (numbers from 1 to 5 in Figure 14.4) and the guard quarter of 8 × 3 m (number 9) at the entrance to the courtyard. The opposite wing has three *contubernia* of similar characteristics (numbers from 6 to 8) and the centurion's suite of rooms (number 10) at the entrance to

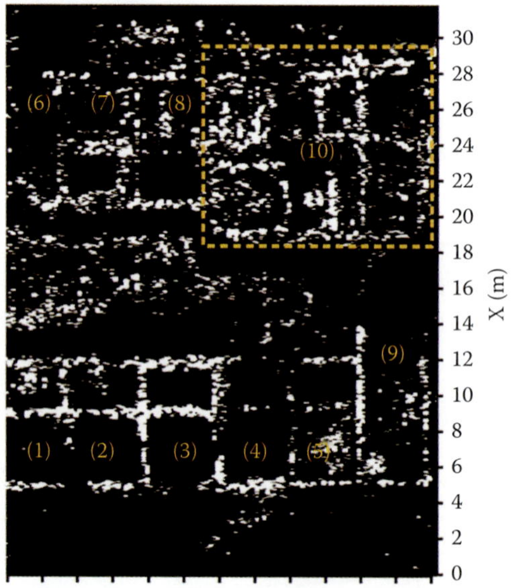

FIGURE 14.4 Results of the Grid 2 at the northern sector of the camp: 3D overlaid image showing the interpretation of the *contubernia* identified (1–8), as well as the guard quarter (9) and the centurion's suite (10).

the courtyard. Unfortunately, it was not possible to prospect the complete structure, and the two opposite *contubernia* at the back of the courtyard and the central cistern were missed in the GPR imaging. It is important to highlight that, although the other excavated barracks present the centurion's suite at the left wing of the entrance (see Figure 14.1), the barrack under this grid has the centurion's suite at the right wing of the entrance.

14.4.3 Visualisation of *Aquis Querquennis*: Integration of TLS and GPR Results

The remaining step of this section involves the geometrical coregistration of ortho-images from different sensors under the same coordinate system. In particular, data integration is performed through the fusion of colour LiDAR and radar imagery obtained over the same area by means of specific ground control points (GCP) placed on-site. For this task, eight GCPs were located in each of the four corners of both rectangular grids surveyed with the GPR technique (see Figure 14.1).

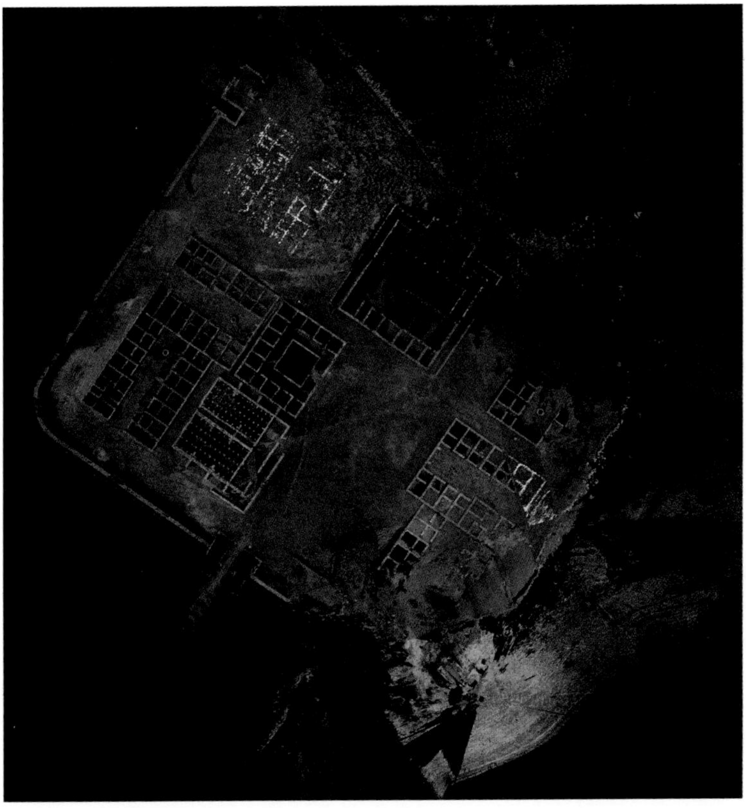

FIGURE 14.5 GPR results produced in Grids 1 and 2 fused on a TLS-based orthoimage of the Roman site.

After registration, consisting of a rigid body transformation where seven parameters are defined (a 3D translation, a 3D rotation and a scaling factor), both GPR grids are integrated under the same local coordinate system defined by the TLS survey (see Figure 14.5). If the point cloud is already georeferenced, the Cartesian representation enables direct retrieval of the coordinates of the points in an absolute coordinate system. If it is not, the data in a local coordinate system may need to be converted into a global coordinate system using, for example, a general purpose simulation system (GPSS) base station.

Finally, it should be noted that since a multidata source approach provides different resolutions, it was possible to generate orthoimages with different spatial resolutions.

14.5 CONCLUSIONS

The application of geomatic techniques in the Roman site *Aquis Querquennis*, recognised as a cultural heritage resource, has widely contributed to its documentation and 3D reconstruction. The integration of TLS and GPR was used for the first time in the encampment to support and complement the visual observations and petrophysical characterisation carried out by previous restorers. The Faro Focus^{3D} X330 has shown its capabilities for measuring the 3D geometry of the overall settlement, letting users obtain, for example, the lengths, heights and volumes of every building or the width of a particular stone pillar, without making direct contact with them. Other interesting LiDAR-derived products, such as 2D plans and orthoimages, can also play an important role for heritage documentation purposes.

Of special interest is the use of GPR results to reveal hidden construction or structural details, such as the right wing of one barrack in Grid 1 or the partial structure of another one in Grid 2, as well as human remains or burial-related materials, which absolutely cannot be observed during visual exploration.

Finally, the new tool 'toGPRi' allowed the generation of precise multisource orthoimages, using the 3D GPR imaging and the orthoimages from TLS, for a deeper analysis of both visible and underground space. At this point, the tool only supports raster layers; however, for future work, the implementation of vectorial layers in order to, for example, compute the volume of buried structures is expected. This volumetric reconstruction will facilitate the dimensioning of the buildings while providing an intuitive and easily understood layout of the underground space distribution.

REFERENCES

Besl, P. and N. McKay. 1992. A method for registration of 3-D shapes. *IEEE Transactions on Pattern Analysis and Machine Intelligence* 14, no. 2 (February): 239–56.

Corsi, C., B. Slapšak and F. Vermeulen. 2013. *Good Practice in Archaeological Diagnostics: Non-Invasive Survey of Complex Archaeological Sites*. Cham, Switzerland: Springer.

Faro Technologies, Inc. http://www.faro.com/ (accessed December 23, 2016).

González-Jorge, H., D. González-Aguilera, P. Rodríguez-Gonzálvez and P. Arias. 2012. Monitoring biological crusts in civil engineering structures using intensity data from terrestrial laser scanners. *Construction and Building Materials* 31: 119–28.

Goodman, D. and S. Piro. 2013. *GPR Remote Sensing in Archaeology, Geotechnologies and the Environment 9*. Heidelberg, Germany: Springer.

Grasmueck, M., R. Weger and H. Horstmeyer. 2005. Full-resolution 3D GPR imaging. *Geophysics* 70: K12–K19.

Leckebusch, J. 2003. Ground-penetrating radar: A modern three-dimensional prospection method. *Archaeological Prospection* 10: 213–40.

Lerma, J. L., S. Navarro, M. Cabrelles and V. Villaverde. 2010. Terrestrial laser scanning and close range photogrammetry for 3D archaeological documentation: The Upper Palaeolithic Cave of Parpalló as a case study. *Journal of Archaeological Science* 37, no. 3 (March): 499–507.

Lin, A., A. Novo, S. Har-Noy, N. Ricklin and K. Stamatiou. 2011. Combining GeoEye-1 satellite remote sensing, UAV aerial imaging, and geophysical surveys in anomaly detection applied to archaeology. *IEEE Journal Selected Topics in Applied Earth Observations and Remote Sensing* 4: 870–76.

Novo, A., M. Solla, J. L. Montero-Fenollós and H. Lorenzo. 2014. Searching for the remains of an early bronze age city at Tell Qubr Abu al-ʻAtiq (Syria) through archaeological investigations and GPR imaging. *Journal of Cultural Heritage* 15: 575–79.

Núñez-Nieto, X., M. Solla, A. Novo and H. Lorenzo. 2014. Three-dimensional ground-penetrating radar methodologies for the characterization and volumetric reconstruction of underground tunneling. *Construction and Building Materials* 71: 551–60.

Puente, I., R. Lindenbergh, H. González-Jorge and P. Arias. 2014b. Terrestrial laser scanning for geometry extraction and change monitoring of rubble mound breakwaters. *ISPRS Annals of the Photogrammetry, Remote Sensing and Spatial Information Sciences* II-5: 289–95.

Puente, I., J. Sande, H. González-Jorge et al. 2014a. Novel image analysis approach to the terrestrial LiDAR monitoring of damage in rubble mound breakwaters. *Ocean Engineering* 91, no. 15 (November): 273–80.

Puente, I., M. Solla, H. González-Jorge and P. Arias. 2015. NDT documentation and evaluation of the Roman Bridge of Lugo using GPR and mobile and static LiDAR. *Journal of Performance of Constructed Facilities* 29, no.1.

Riveiro, B., P. Morer, P. Arias and I. de Arteaga. 2011. Terrestrial laser scanning and limit analysis of masonry arch bridges. *Construction and Building Materials* 25, no. 4 (April): 1726–35.

Rodríguez Colmenero, A. 2016. *Aquae Querquennae, Roman Headquarters and Inn: A Guide to the Archaelogical Site*. Dirección Xeral de Turismo: Xunta de Galicia.

Vecco, M. 2010. A definition of cultural heritage: From the tangible to the intangible. *Journal of Cultural Heritage* 11, no. 3 (July–September): 321–24.

Yastikli, N. 2007. Documentation of cultural heritage using digital photogrammetry and laser scanning. *Journal of Cultural Heritage* 8, no. 4 (September–December): 423–27.

15 Earthquakes, Tsunamis and Harbours

A Geoarchaeological GPR-Based Approach for Depicting the Spatial Characteristics of Human Structures and Natural Hazards in Ancient Falasarna Harbour

Michael Styllas and Klisthenis Dimitriadis

CONTENTS

15.1 INTRODUCTION

In the broader context of geoarchaeological research, the discovery of ancient settlements through archaeological excavations often brings up significant archaeological and geological information. The knowledge obtained during excavations not only provides explanations of the relationship of past societies with the evolving natural environment, but can also offer alternative scenarios about the likely causes and conditions that may have forced the abandonment of antique and/or historical settlements. In the case where ancient urban complexes are located in coastal areas, the discovery of structures, such as harbours and jetties, provides additional information concerning the alluviation history and other environmental processes. Additionally, when such structures have been displaced by the force of seismic events or even when they have been covered by tsunami deposits, a wealth of geophysical and archaeological information is revealed once those man-made constructions are excavated. Excavations are costly though, and in many cases, move blindly, so the use of nondestructive geophysical methods such as ground-penetrating radar (GPR) and electrical resistivity tomography (ERT) becomes a real asset.

In this chapter, we present the application of GPR in the ancient harbour of Falasarna in north-western Crete (Figure 15.1a). Even though the Ancient Falasarna archaeological site includes a series of man-made constructions, such as bridges, towers and water tanks, here we present only the results from the vicinity of the harbour. GPR proved to be very efficient in tracing the lateral extent both of the harbour's main jetty and of the tsunami sediments, deposited in the area during two major historical seismic events.

15.2 REGIONAL SETTING AND RESEARCH FRAMEWORK

From a geophysical standpoint, Ancient Falasarna is positioned within an area of intense tectonism in an extensional (normal) fault domain on the back of the Outer Hellenic Arc, along which the African plate is subducting under the Eurasian plate (Le Pichon and Angelier, 1981; Lallemant et al., 1994). The area to the west of Crete, including Ancient Falasarna, is characterised by inland and offshore fault systems that trend N–S (Figure 15.1a).

The surficial traces of active faults and of the Holocene tectonic uplift are recognisable in the Falasarna harbour area (Figure 15.1b), which, in the past, has been affected by both relatively shallow earthquakes associated with extensional faulting and by deeper earthquakes associated with the thrust movement of the subducting plate (Dominey-Howes et al., 1998). There is a long history of seismicity and tsunami action in the eastern Mediterranean and especially in western Crete (Scheffers and Scheffers, 2007). The most extreme neotectonic event in historical times occurred in AD 365. It had an estimated magnitude of $Ms = 8.3$ R, while the co-seismic uplift during the earthquake was 9 m (Thommeret et al., 1981; Pirazzoli et al., 1996; Stiros, 2001; Stiros and Papageorgiou, 2001). In the Ancient Falasarna harbour, this is mostly apparent not only from the up-and-inland displacement of the harbour jetties, but also from the late Holocene sedimentation record that includes layers of mixed deposits of sandy material and large flattened clasts, transported to the area by successive tsunamis triggered during the AD 66 and AD 365 earthquakes (Pirazzoli et al., 1992). The archaeological and palaeoenvironmental history of Falasarna has been investigated along five trenches, which were excavated by the Greek Regional Archaeological Service. The goal was twofold: to define the lateral extent of harbour jetties and of other constructions, and to further study the sedimentary sequences associated with those high magnitude seismic events. The local stratigraphy as revealed from the excavated trench in the harbour area consists of the following units:

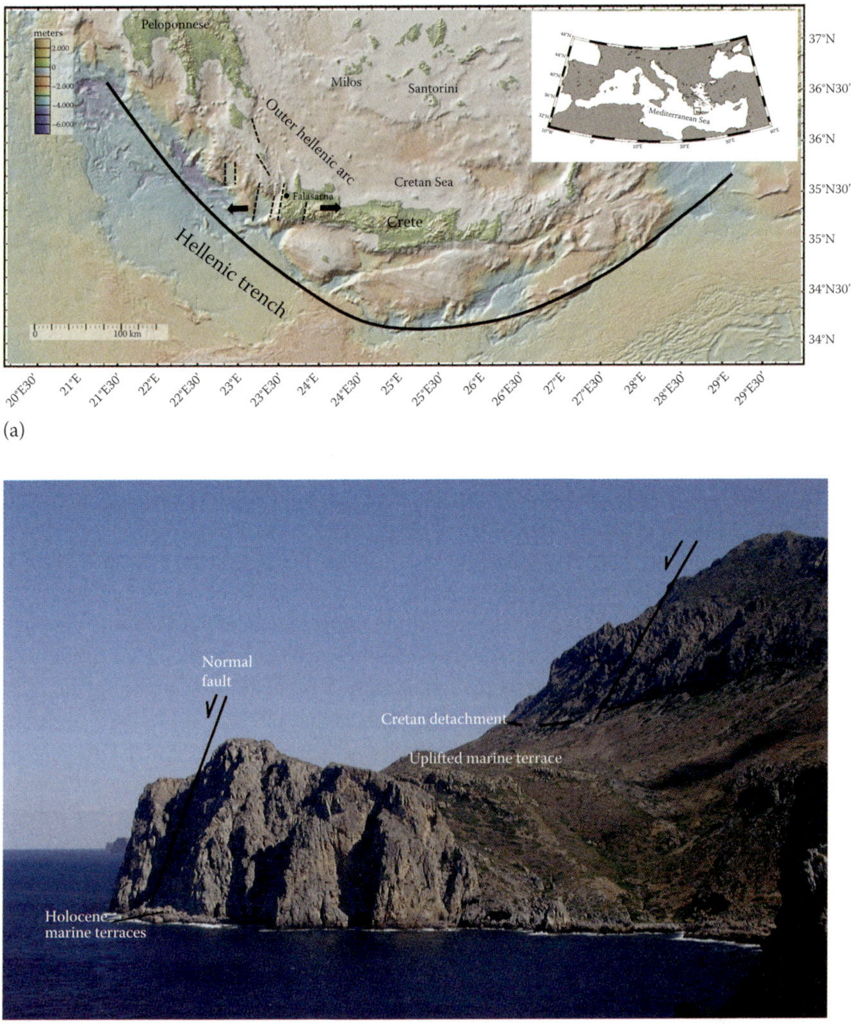

FIGURE 15.1 (a) General setting of Ancient Falasarna in relation to the tectonic extensional field (black arrows) of the Outer Hellenic arc. Dotted black lines correspond to seismically active major normal faults with respective N–S orientation. (Topographic/bathymetric background from http://www.geomapapp.org/.) (b) View looking to the north from the Ancient Falasarna harbour. Detachment (normal) active faults counteracting with the uplift of western Crete as indicated by the existence of marine terraces, the youngest of which date from the Holocene epoch.

- Alluvial land deposits. A fine-grained (silt/clay) layer of gravel and cobbles not exceeding 30 cm in thickness.
- Tsunami deposits. A layer of variable thicknesses, with marine sandy material and flat clasts of nearshore origin.

- Marine deposits. Mainly fossiliferous sandy material reworked by the waves which comprise the harbour sedimentary fill.
- Limestone bedrock. Fractured and partly saturated with seawater located 6 m below the surface.

The preexisting knowledge about the stratification of surface sediment layers and the relatively flat terrain in the vicinity of the archaeological site permitted a series of geophysical nondestructive applications to be applied with a specific focus on the extension of the harbour jetty and the spatial distribution of the tsunami deposits near the harbour.

15.3 METHODOLOGY, EQUIPMENT AND GPR CALIBRATION

GPR is a nondestructive, fast, low-cost and precise investigation technique that uses reflected electromagnetic (EM) waves. In modern archaeology, nondestructive geophysical site assessment methods are of great importance as they are used to provide a fast and accurate picture of subsurface conditions, prior to any excavation works. Contemporary to its archaeological applications, GPR has shown its potential for detecting stratigraphic architecture, layer boundaries and sand-body geometry (e.g. Koster et al., 2014 and references therein). A dense grid of GPR data allows for analyses of the lateral extent of specific sediment for relatively large areas to be quickly undertaken. During tsunami deposit surveys, a shallow groundwater table and/or seawater intrusions may limit GPR measurements. This is because high electrical conductivities reduce penetration depths and data quality. More limiting factors include uneven and/or rough surface conditions, changes in lithology and soil humidity, anthropogenic modification of the subsurface and the presence of metallic objects or other radar wave sources. Fine grain sizes can influence the resolution of GPR, for example high clay content in subsurface layers can lead to attenuation of the EM waves. In the Ancient Falasarna archaeological site, field methodology involved a two-step procedure. An initial coverage of an irregular grid of GPR profiles was carried out to discern the broad extent of tsunami deposits and locate sub-areas of archaeological interest, where a more intensive research is required, followed by denser grids along the sub-areas to map all human-made structures in the vicinity of the harbour with the best possible accuracy.

For accurate geoarchaeological assessments, detailed topographical information is required prior to any geophysical profiling, to assign the required topographical information to all findings. This was achieved by using a differential global positioning system (DGPS) to create a set of tie points in the area with centimetre precision, so that all geophysical profiles and derived surfaces are 'tied' in these known points. Topographical background information maps were provided by the Greek Military Geographical Service (GYS) and digitised in a WGS84 geographical system with a Universal Transverse Mercator (UTM) coordinate format. To minimise errors, two types of global positioning systems (GPS) were used along with a correction protocol: (i) a Garmin E-trex GPS for coarse gridding (accuracy 6–7 m); (ii) a Trimble Geoexplorer, Series 2003, DGPS (SBAS/EGNOS with augmented accuracy reaching 0.5 m); and (iii) DGPS corrections, using the Radio Technical Commission for Maritime Services (RTCM) protocol, retrieved from a global system for mobile communications (GSM) Internet link to the permanent station of the JGC company and fed to the GARMIN GPS in real time.

The only available archaeological map of the Ancient Falasarna archaeological site was provided by Elpida Hatzidaki, senior archaeologist of the Falasarna site (pers. comm.), and was orthorectified with the use of the network tie points and inserted into a geographic information system (GIS) platform. In addition, the mapping of other ruins using a GPS with differential correction (RTCM correction) led to the construction of the first complete orthorectified archaeological map of Ancient Falasarna that was used as background for GPR surveys.

Currently, all available information on the Falasarna site comes from a few square metres of excavations made along the excavated trenches. The shape of the harbour and its surroundings is vaguely known and the rest of the ancient city probably extends inland. To detect subsurface human constructions, GPR scans were realised in selected sub-areas where surface topography denoted the existence of subsurface constructions. The smooth relief of the surficial alluvial cover of variable thicknesses permitted the efficient application of GPR. To ensure continuous movement, the shrub vegetation was cleared and the following sub-areas were measured in an orthorectangular grid, including the main harbour entrance, the main harbour central part, the round tower area, the Dexameni ('Δεξαμενή'—water tank in Greek) area, the second harbour, the lake area, the area of Tafros ('Τάφρος'—trench in Greek) and the north walls (Figure 15.3). GPR surveying was realised with the following equipment: MCII – RAMAC GPR, main unit, 500 and 250 MHz shielded antennas and a 25 MHz (rough terrain concept) antenna. The interpretation of the derived GPR data was made with Object Mapper software from MALA GEOSCIENCE. The software composes the plan view using a combination of multiple GPR profiles (radargrams), as shown in Figure 15.2.

15.4 RESULTS: GPR MAPPING OF ANCIENT FALASARNA HARBOUR

The main objectives of GPR profiling were to map the extent and geometrical characteristics of alluvial and tsunami deposits and also the subsurface human-made utilities (walls, jetties) in the harbour area. Grid spacing between the GPR profiles was 2 m. The surficial alluvial cover demonstrated very low resistivity values due to high clay content and resulted in considerable attenuation of EM antenna pulses, decreasing the penetration depth. In addition to high concentrations of fine-grained material, the surface layer also contains a mixture of medium-sized cobbles, likely transported from the mountainous area in the vicinity of the harbour. These cobbles are sometimes identifiable in the GPR radargrams, but at other times produce diffraction effects, distorting the general subsurface picture. Antenna ringing also occurred at some points due to the highly conductive ground. Due to these site-specific restrictions, only scans with the 500 MHz antenna were realised and considered hereafter.

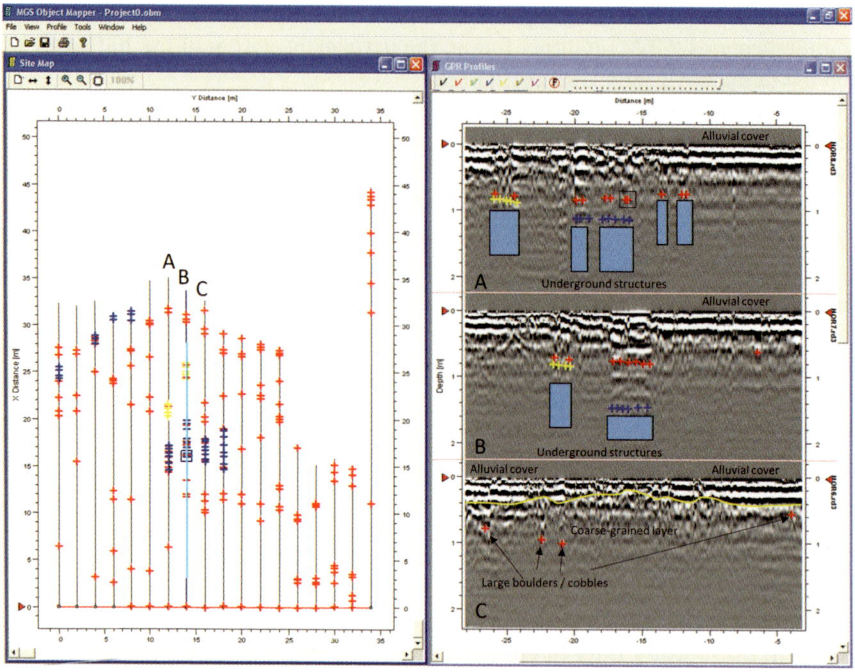

FIGURE 15.2 Three GPR profiles (radargrams) corresponding to profiles A, B and C, shown in the interpretation screen to the right. The horizontal red line is the surveyed grid baseline (distance in m). Profiles are along the x-axis (black lines) and the coloured crosses represent different types of findings, allocated at a specific depth of a specific profile and likely correspond to human structures (right panel). Explanatory radargram C reveals the existence of coarse clasts below the surface alluvial cover (yellow line), potentially related to tsunami deposits.

15.4.1 Main Harbour

The GPR application in the harbour area succeeded in identifying stratigraphic Layers 1–3 that correspond to alluvial, tsunami and marine deposits (Figure 15.3, purple crosses). Horizon 4 (bedrock) was not reached by GPR in the harbour area. Based on the principle of continuity, the absence of any of the foregoing layers with sharp boundaries is interpreted here as human-made structures, such as walls and/or harbour jetties (Figures 15.3 and 15.4, yellow crosses). Strong reflections from greater depths are likely to correspond to other man-made structures (Figures 15.3 and 15.4, red crosses) as well. The extent of alluvial deposits is easily identified in all radargrams and mapped into the GIS system. The thicknesses of these deposits are uniform, ranging between 20 and 30 cm, without significant lateral variations towards the harbour.

The underlying layer (tsunami deposits) also appears uniform with considerable variations at the harbour sides. The shape of the surface alluvial fine-grained mixed layer with the very low and locally minimal gradients is explained by the hydrogeological properties of the harbour sedimentary fill. In fact, the flat topography of

FIGURE 15.3 Perspective view of the main GPR grid in the central harbour area. Yellow spots denote the discontinuity in the thin alluvial superficial deposits, mainly due to the extension of the harbour jetties towards the north-west. Dotted black line (A–B) denotes the location of a long (380 m) GPR transect, presented in Figure 15.5c.

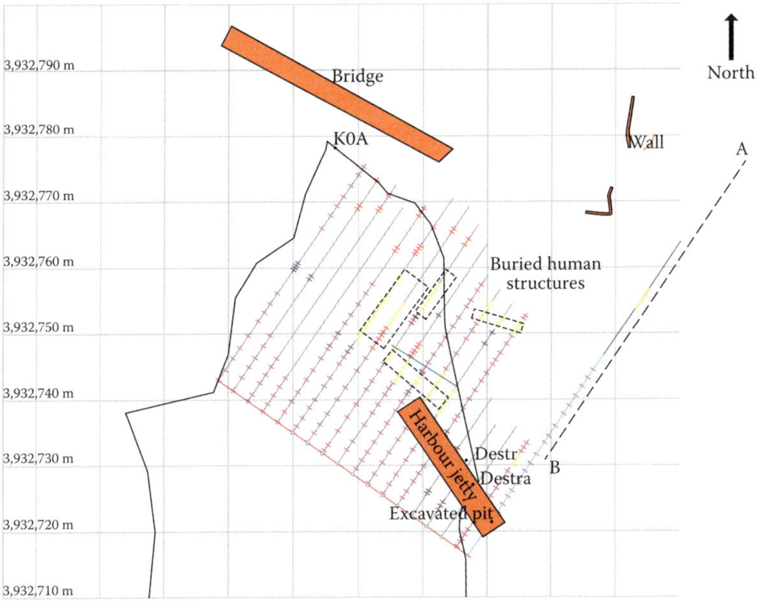

FIGURE 15.4 Interpretation of the central harbour GPR grid. Purple crosses denote the existence of a relatively thick (20–30 cm) clay cover. Yellow crosses denote the discontinuity of the surface cover, probably due to the extension of the harbour jetties or of other human constructions (dashed-lined boxes) in the ancient harbour towards the north-east.

the harbour area and the medium to high permeability of the underlying layers (tsunami and marine deposits) transform the harbour into a temporary lake during rainy periods. This stratigraphic setting results in temporary pond-like conditions with concurrent deposition of fine material (silts, clays), derived either from the catchment and/or from direct aeolian deposition, as happens during incidents of intense dust transported from the North Africa (Sahara) region, which is known to influence eastern Mediterranean sedimentation from mid-Holocene onwards (Ehrmann et al., 2007). The two-dimensional representations of yellow-coloured areas on the site (Figure 15.4) denote the discontinuity between the first and second layer that exists in some profiles to the north-west of the main excavated jetty. Thus, we can draw the conclusion that the existing jetty also extends to the north-west for some more metres.

15.4.2 Tsunami Layer

Figure 15.5a illustrates the excavated trench near the harbour jetty with the two characteristic boat moorings – Destra ('Δέστρα' – boat moorings in Greek) – along with the sediment profile revealed during excavation. The tsunami deposits are visible beneath the surface of a fine-grained layer (10–20 cm), with stacked flattened cobbles and variable thicknesses.

The coarse gravel contained in the surface layer locally, created strong hyperbolic reflections. The contact between the superficial and the tsunami layers was successfully mapped by GPR using the 500 MHz antenna (Figure 15.5b and c). Despite the surface layer of fine deposits that resulted in high attenuation of the 500 MHz antenna, the sandy and coarse clast deposits were characterised by dozens of hyperbolae below the surface layer (represented by the yellow line in Figure 15.5b and c), with velocities of ~8–16 m/ns, which reflect the high gravel and boulder content. The depth of the tsunami layer fits well with the sedimentological evidence.

Following the GPR calibration procedure, additional scans were realised in the vicinity of the harbour to map the lateral extent and the respective thicknesses of tsunami deposits. In other GPR profiles within the Falasarna harbour, the tsunami layer was vertically more extensive, with a respective thickness of 30–50 cm. These GPR results were verified in 2008 when excavations of new trenches and pits within the ancient harbour revealed a thick layer of coarse gravel and marine cobbles above 1.5 m of marine sandy deposits. Away from the harbour jetty, tsunami deposits had a thickness of up to 1.5 m. This may have resulted from wave refraction effects caused by the jetty itself, when the tsunami struck Ancient Falasarna. In most of the GPR profiles, the attenuation of the EM waves rises near the centre of the harbour, which may be due to the higher clay content on the surface. This again outlines the importance of the sedimentological setting of the ground surface in GPR study areas.

15.5 CONCLUSIONS

Our GPR-based geophysical investigations reveal new insights about the subsurface man-made structures and inferred tsunami deposits in the Ancient Falasarna

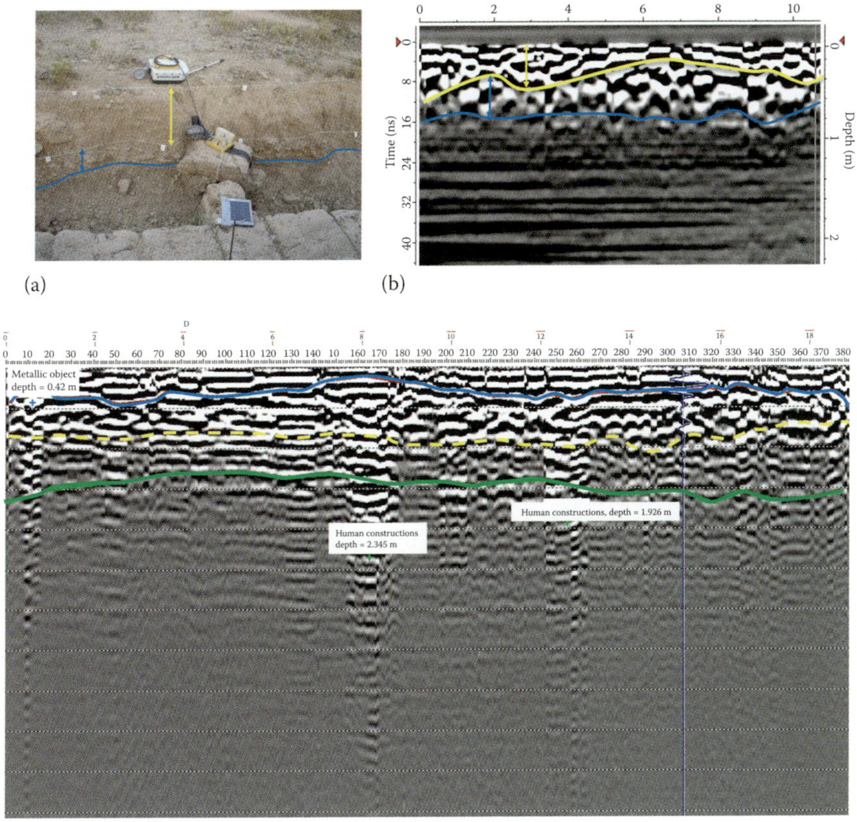

(a) (b) (c)

FIGURE 15.5 (a) Excavated trench along the harbour jetty including the stratigraphic horizons of the surface alluvial layer (yellow arrow), tsunami deposits (blue arrow) and the underlying marine fossiliferous deposits. (b) Processed GPR radargrams with two-way traveltime (TWT) and depth in metres that led to the respective calibration with the local sedimentology as exposed in the trench. (c) One of the longest (380 m) radargrams (A–B in Figures 15.4 and 15.5) realised away from the harbour jetty. The thickness of the stacked tsunami deposits is considerably larger (40–1.5 m) compared with the layers of the excavated harbour trench (30–50 cm).

harbour, along the north-western coast of Crete. Based on excavated trench sedimentological evidence, a surface layer of fine-grained (silt and clay) deposits containing angular gravel fragments and terrestrial molluscs overlies a distinct layer containing a mixture of gravel, sand, shell fragments and flattened clasts that represent the tsunami deposits, triggered by the AD 66 and in all likelihood by the AD 385 earthquakes too. An underlying mixed layer, mainly consisting of marine sands with marine fauna, was what the harbour was filled with prior to these major seismic events. GPR data were successful in mapping the lateral extents and the extension of the ancient harbour main jetty, where ship moorings are still visible, as well as other man-made constructions throughout the archaeological site of Ancient Falasarna. These features were verified using a noninvasive geophysical investigation method such as GPR. Without GPR large-scale trenching, these investigations cannot be carried out. The lateral extent of the tsunami deposits appear to be of a greater thickness away from the jetty, a probable result of the tsunami wave refraction at the margins of the jetty. Detailed knowledge of the spatial distribution of tsunami deposits, which was beyond the scope of the present work, together with knowledge of the grain-size characteristics of the transported clasts, as derived by the archaeological trenches, can be used in the future as input to inverse numerical models, in order to reconstruct the characteristics of the tsunami wave (e.g. significant height and period) and of the earthquake's magnitude as well.

REFERENCES

Dominey-Howes, T. M., A. Dawson, and D. Smith. 1998. Late holocene coastal tectonics at Falasarna, western Crete: A sedimentary study. *Geological Society of London, Special Publications* 146: 343–352.

Ehrmann, W., G. Schmiedl, Y. Hamman, et al. 2007. Clay minerals in late glacial and holocene sediments in northern and southern Aegean Sea. *Palaeogeography Palaeoclimatology Palaeoecology* 249: 36–57.

Koster, B., G. Hoffmann, C. Grützner, and K. Reicherter. 2014. Ground penetrating radar facies of inferred tsunami deposits on the shores of the Arabian Sea (northern Indian Ocean). *Marine Geology* 351: 13–24.

Lallemant, S., C. Truffert, L. Jovilet, P. Henry, N. Chamot-Rooke, and B. De Voogd. 1994. Spatial transition from compression to transition in the western Mediterranean Ridge accretionary complex. *Tectonophysics* 234: 33–52.

Le Pichon, X., and J. Angelier. 1981. The Aegean Sea. *Philosophical Transactions of the Royal Society of London, Series A* 300: 357–372.

Pirazzoli, P. A., J. Ausseil-Badie, P. Giresse, E. Hatzidaki, and M. Arnold. 1992. Historical environmental changes at Phalasarna harbor, West Crete. *Geoarchaeology* 7: 371–392.

Pirazzoli, P. A., J. Laborel, and S. C. Stiros. 1996. Coastal indicators of rapid uplift and subsidence: Examples from Crete and other Mediterranean sites. *Zeitschrift für Geomorphologie* NF Suppl.-Bd. 102: 21–35.

Scheffers, A., and S. Scheffers. 2007. Tsunami deposits on the coastline of west Crete (Greece). *Earth and Planetary Science Letters* 259: 613–624.

Stiros, S. C., 2001. The 365AD Crete earthquake and possible seismic clustering during the 4–6th centuries AD in the eastern Mediterranean: A review of historical and archaeological data. *Journal of Structural Geology* 23: 545–562.

Stiros, S. C., and S. Papageorgiou. 2001. Seismicity of western Crete and the destruction of the town of Kisamos at AD 365: Archaeological evidence. *Journal of Seismology* 5: 381–397.

Thommeret, Y., J. Thommeret, J. Laborel, L. F. Montaggioni, and P. A. Pirazzoli. 1981. Late holocene shoreline changes and seismotectonic displacements in western Crete (Greece). *Zeitschrift für Geomorphologie* Supplementband 40: 127–149.

16 Ground-Penetrating Radar for Nondestructive Structural Imaging of Ancient Sculptures

Klisthenis Dimitriadis, Georgianna Moraitou, Michael Styllas, Ioannis Panagakos, Ioannis Konstantakis and Sofia Avlonitou

CONTENTS

16.1 INTRODUCTION

The collection of marble statues from the Antikythera Shipwreck consists of numerous statues sculpted in white marble from the Greek Island of Paros. The statues were discovered by fishermen on the seabed off the Island of Antikythera, located south of Peloponnese in 1900. The statues were found in a fragmentary condition and they were conserved either immediately after they became part of the National Archaeological Museum's collections or later in the 1970s. They were restored by adhering the broken parts and installing metallic dowels within the joints. Unfortunately, the conservation records are missing and hence there is no additional information concerning the emplacement or the size of each bar in these statues. From September 2015 to March 2016, the successful temporary exhibition of the National Archaeological Museum of Athens titled 'The Antikythera Shipwreck: The Ship, the Treasures, the Mechanism' was to be exhibited at the Antikenmuseum Basel und Sammlung Ludwig in Switzerland. For this reason, the static behaviour of five very important marble statues from the Antikythera collection had to be investigated, prior to 'securely packing' them for long-distance transport.

In order to ensure that metallic reinforcement was used in the joints, it was necessary to image the nonvisible metallic dowels installed in these statues. The prime option was to use γ-ray imaging, which would give a positive image of the dowels with all the details of their dimensions, surface relief, and so on. The disadvantage of γ-ray imaging is that it is expensive and requires high levels of protection. So, it was decided to use instead the ground-penetrating radar (GPR) technique that was believed to give the second-best results. The task of implementing the GPR technique was allocated to GEOSERVICE Ltd. The aim of the project was to locate and if possible 'map' the geometry of the nonvisible metallic dowels installed in the statues, using the GPR technique.

To achieve this task, GEOSERVICE Ltd., in collaboration with the Department of Conservation, Physical, Chemical Research & Archaeometry of the National Archaeological Museum developed a new method adapted to statues (Dimitriadis, 2015). The method was applied in five specific statues (Vlachogianni, 2015):

1. Statue of a boy (wrestler?) (inv. Nr. 2773)
2. Statue of Hermes (inv. Nr. 2774)
3. Statue of a helmeted warrior (inv. Nr. 15534)
4. Statue of Ulysses (inv. Nr. 5745)
5. Statue of a man (Achilles?) (inv. Nr. 5746)

16.2 GROUND-PENETRATING RADAR METHOD

The GPR method is a fast, noninvasive, low-cost and precise investigation technique that uses reflected electromagnetic waves (e.g. Bristow and Jol, 2003; Neal, 2004). The GPR technique was invented by the U.S. military forces during the Vietnam War to locate mines. Later on, the method was developed to serve the geophysics sector and map any kind of object (metallic and nonmetallic) below the ground surface. The GPR method is based on one very simple concept: One short electromagnetic pulse is emitted into the ground. If the ground is fully uniform, the pulse travels through it and nothing comes up as a reflection from one possible ground discontinuity. On the contrary, if there is something below the ground with different electrical characteristics from the surrounding environment, then part of the emitted energy is reflected back and part of it travels to deeper layers. This simple physical concept applies to all existing earth media and sediments such as limestones, sandstones, clay and sand layers as well as other types of rock formations. By using this concept and specific instruments, it is possible to investigate the interior of objects made from the foregoing materials.

Over the past decades, the GPR method has obtained widespread status with applications in civil engineering (e.g. Hugenschmidt, 2002; Barnes et al., 2008; Lahouar and Al-Qadi, 2008), archaeology (e.g. Negri and Leucci, 2006; Perez-Gracia et al., 2009), surface geology (e.g. Neal and Roberts, 2001) as well as in the mapping of tsunami deposits (e.g. Koster et al., 2014)

16.3 INSTRUMENTATION AND SCANNING METHODOLOGY

In this project, a modern GPR system, the CX-12 system of MALA GEOSCIENCE, was used. The system consists of one fully integrated GPR main unit, fitted with a

FIGURE 16.1 The presence of joints on the statue of the 'Wrestler of Antikythera' indicating the possible existence of metallic dowels. (Photo: National Archaeological Museum of Athens/Archaeological Receipts Fund.)

liquid-crystal display (LCD) screen and operated with high-frequency antennas of 1500 and 2300 MHz. For the purpose of this project, we used a 2300 MHz antenna, which provided a resolution accuracy of 0.5 cm and a penetration depth of 50 cm within the selected material.

In all cases, the material was marble. The geophysical targets were the metallic bars in the marble of each statue. Usually these bars (dowels) are located at the centre of broken surfaces, and connect the broken parts of the statue. Their existence is evident from the visible cracks at specific points of each statue (Figure 16.1).

The scanning process was based on performing a two-dimensional (2D) grid of parallel scans with a grid size (cell) of 5 cm. The *X* and *Y* sizes of the grid varied according to the specific area to be investigated. For each target, one separate grid was performed.

The data processing was realised using the dedicated MALA GEOSCIENCE software, '3D Vision', licensed to GEOSERVICE Ltd. In general, metallic targets appear as strong white areas (reflections) against a black/grey background. In many cases, the parallel use of one metal detector in addition to one other physical method proved very effective to validate the existence of metal. However, the metal detector was unable to give the geometry of the metal and it was used only in the binary concept of Y/N for the existence of metal.

Figure 16.2 illustrates the scanning process of one specific area of the 'Wrestler of Antikythera' statue.

16.4 VELOCITY CALIBRATION PROCESS

Although the application of GPR in a 2D space is a relatively easy and widely used procedure, especially with cement surfaces, in the case of statues with their three-dimensional geometry, it becomes a very hard and challenging process. The reason is because the emitted pulse travels in nonuniform media, composed of air. Marble,

FIGURE 16.2 The process of grid-scanning one specific area of the 'Wrestler of Antikythera' statue. (Photo: National Archaeological Museum of Athens/Archaeological Receipts Fund/GEOSERVICE.)

due to its surface irregularity and the resulting reflections, does not correspond to the exact depth where they are located. The velocity of the electromagnetic pulse in the air is known (30 cm/ns) and in the case of marble it varies from 10 to 12 cm/ns.

For this reason, a specific velocity calibration was carried out on one flat surface of white marble and the measured/calculated velocity was found to be 10.2 cm/ns.

16.5 RESULTS: CASE STUDY STATUE 'WRESTLER OF ANTIKYTHERA'

GPR gridded scans were realised in five statues (Section 16.1) but here we present only the results of the 'Wrestler of Antikythera'. The 'Wrestler of Antikythera' statue was found partially buried on the seabed. Part of the statue (its left side) was exposed to sea erosion and sea boring organisms for more than 2000 years, which contributed to significant erosion of the construction material. On the contrary, the right side of the statue, which was buried below the seabed and in the sand, was intact as the sea microorganisms were unable to affect that side. In addition to being eroded, the statue was found broken into four pieces.

FIGURE 16.3 Superposition of the raster image from the GPR scan on the 'Wrestler of Antikythera' Statue. (Photo: National Archaeological Museum of Athens/Archaeological Receipts Fund/GEOSERVICE.)

According to the visible cracks, three specific grids were installed to examine the statue. The result of the raster output from the GPR scan showed the emplacement of the internal metal bars (Figure 16.3). In fact, the metal acted as a strong reflector and produced a whitish area in the grid slices. By replacing the plexiglass grid with one virtual graphics grid and adjusting the images, the location and the extent of each metal bar were mapped on the statue. Using this form of visualisation, the following results were derived from GPR measurements.

Separate scans were performed on all sides of the 'Wrestler of Antikythera' statue, and the processed grids were very accurate in depicting the exact position of the metallic bars installed during restoration, emphasising the applicability and efficiency of the method presented here (Figure 16.4).

16.6 CONCLUSIONS

The innovative method presented here can be easily applied to numerous cases of ancient sculptures, accurately describing the geometry of metallic dowels placed during restoration. Prior knowledge of the geometrical characteristics of the restoration interventions, when no documentation is available, can protect ancient statues as well as other marble constructions.

FIGURE 16.4 The 'Wrestler of Antikythera' statue with the subsequent GPR grid, showing areas with reinforcing metal bars. The bars are mapped as white reflections and graphically shown in black. (Photo: National Archaeological Museum of Athens/Archaeological Receipts Fund/GEOSERVICE.)

16.7 ACKNOWLEDGEMENT

The project was funded by Antikenmuseum Basel und Sammlung Ludwig.

REFERENCES

Barnes, C. L., Trottier, J.-F., Forgeron, D., 2008. Improved concrete bridge deck evaluation 11 using GPR by accounting for signal depth-amplitude effects. *NDT & E International*, 41, 427–433.

Bristow, C. S. and Jol, H. M., 2003. An introduction to ground penetrating radar (GPR) in sediments. *Geological Society*, London, Special Publications 211 (1), 1–7.

Dimitriadis, K., Konstantakis, I., Avlonitou, S., Panagakos, I. and Moraitou, G., 2016. Ground penetrating radar for non-destructive structural imaging of ancient sculpture, in *41st International Symposium on Archaeometry (ISA), Book of Abstracts*, N. Zacharias and E. Palamara (eds), Kalamata, Greece, May 15–21, University of Peloponnese, p. 204.

Hugenschmidt, J., 2002. Concrete bridge inspection with a mobile GPR system. *Construction and Building Materials*, 16, 147–154.

Koster, B., Hoffmann, G., Grützner, C. and Reicherter, K., 2014. Ground penetrating radar facies of inferred tsunami deposits on the shores of the Arabian Sea (Northern Indian Ocean). *Marine Geology*, 351, 13–24.

Lahouar, S. and Al-Qadi, I. L., 2008. Automatic detection of multiple pavement layers from GPR data. *NDT & E International*, 41, 69–81.

Neal, A., 2004. Ground-penetrating radar and its use in sedimentology: Principles, problems and progress. *Earth-Science Reviews*, 66 (3–4), 261–330.

Neal, A. and Roberts, C. L., 2001. Internal structure of a through blowout, determined from migrated ground-penetrating radar profiles. *Sedimentology*, 48, 791–810.

Negri, S. and Leucci, G., 2006. Geophysical investigation of the temple of Apollo (Hierapolis, Turkey). *Journal of Archaeological Science*, 33, 1505–1513.

Perez-Gracia, V., Caselles, O., Clapes, J., Osorio, R., Martines, G. and Canas, J. A., 2009. Integrated near surface geophysical survey of the Cathedral of Mallorca. *Journal of Archaeological Science*, 36, 1289–1299.

Vlachogianni, E., Sculpture, Gods and Heroes from the depths of the sea, in *The Antikythera Shipwreck: The Ship, the Treasures, the Mechanism*, N. Kaltsas, E. Vlachogianni and P. Bouyia (eds), *National Archaeological Museum, Antikenmuseum Basel und Sammlung Ludwig*, 2nd ed., 2015, pp. 78–95.

Index